MANUFACTURED LIGHT

RECORD COPY
Do not remove from office

Edited by
Emiliano Gallaga M. and Marc G. Blainey

MANUFACTURED
LIGHT
MIRRORS IN THE
MESOAMERICAN REALM

UNIVERSITY PRESS OF COLORADO
Boulder

© 2016 by University Press of Colorado

Published by University Press of Colorado
242 Century Circle, Suite 202
Louisville, Colorado 80027

First paperback edition 2019
Printed in the United States of America

 The University Press of Colorado is a proud member of
the Association of University Presses.
ASSOCIATION of UNIVERSITY PRESSES

The University Press of Colorado is a cooperative publishing enterprise supported, in part, by Adams State University, Colorado State University, Fort Lewis College, Metropolitan State University of Denver, University of Colorado, University of Northern Colorado, Utah State University, and Western State Colorado University.

∞ This paper meets the requirements of the ANSI/NISO Z39.48-1992 (Permanence of Paper).

ISBN: 978-1-60732-407-2 (cloth)
ISBN: 978-1-60732-938-1 (paperback)
ISBN: 978-1-60732-408-9 (ebook)

Library of Congress Cataloging-in-Publication Data
Manufactured light : mirrors in the Mesoamerican realm / a volume edited by Dr. Emiliano
    Gallaga M., Dr. Marc G. Blainey.
      pages cm
    ISBN 978-1-60732-407-2 (cloth : alkaline paper) — ISBN 978-1-60732-938-1 (pbk: alk.
    paper) — ISBN 978-1-60732-408-9 (ebook)
  1. Indians of Mexico—Antiquities. 2. Indians of North America—Southwest, New—
    Antiquities. 3. Indians of Mexico—Material culture. 4. Indians of North America—
    Southwest, New—Material culture. 5. Mirrors—Mexico—History—To 1500. 6. Mirrors—
    Southwest, New—History—To 1500. 7. Material culture—Mexico—History—To 1500.
    8. Material culture—Southwest, New—History—To 1500. 9. Mexico—Antiquities.
    10. Southwest, New—Antiquities. I. Gallaga, Emiliano, 1970– II. Blainey, Marc Gordon.
  F1219.3.M42M36 2015
  972'.01—dc23

                              2015004618

Cover photograph: Mosaic iron pyrite, National Museum of the American Indian, Smithsonian Institution (14/7000), photo by NMAI photo services.

To Professor John J. McGraw (1974–2016)

pioneer of cognitive anthropology and the
study of Mesoamerican geo-spirituality

Contents

Tables

MANUFACTURED LIGHT

1

"Here is the Mirror of Galadriel," she said. . . .

. . . "What shall we look for, and what shall
we see?" asked Frodo. . . .

. . ."[T]he mirror will also show things unbid-
den, and those are often stranger and more
profitable than things which we wish to behold.
What you will see, if you leave the Mirror free
to work, I cannot tell. For it shows things that
were, and things that are, and things that yet
may be. But which it is that he sees, even the
wisest cannot always tell. Do you wish to look?"

(Tolkien 1991: 381)

Introduction

Emiliano Gallaga M.

In our daily life, it is not a surprise to see our reflection
in a mirror early in the morning and identify that it is
our image reproduced by this solid, reflective surface.
For most people, one's reflection in a mirror is unre-
markable, as we do not attribute a divine quality to seeing our double image. However, while this daily act is
mundane for most of us today, reflected images were
viewed as quite profound by many ancient humans
around the globe, and by pre-Hispanic indigenous
people in particular.

Since the beginning of time, humans have been so
mesmerized and/or challenged by their physical environment that there has always been a need to understand it, to own it, and to transform it. This need applies
not only to our surroundings but to ourselves as well.
We like to know who and what we are, change the way

DOI: 10.5876/9781607324089.c001

3

we look and the things we own, and to make or acquire things that say something about us and about the community to which we belong. This need for knowledge and transformation is an essential spark for the cultural development of the human animal, creating a universe of objects that help us understand and change our environment into a familiar landscape. Among that great universe of items, mirrors or reflecting surfaces have occupied an important place in the human mind. Pendergrast (2003: 13) states that "the ability to recognize themselves in the mirror seems peculiar to superior primates." Humans are likewise captivated by the reproduction of one's own image in a mirror or other reflecting surface. Accordingly, the ancient Indus, Chinese, Egyptian, Greek, Roman, Inca, Aztec, and Maya civilizations created objects that fulfill the need to have and control reflective surfaces (Albenda 1985; Baboula 2000; Beasley 1949; Bulling 1960; Cameron 1979; Cammann 1949; Lilyquist 1979; Pendergrast 2003). Of course, only the gods would know exactly what the ancients would think about the parallel worlds glimpsed through the shiny surfaces of mirrors, an archaeological mystery about which we can now only make educated guesses.

Complex and time-consuming to produce, mirrors and other reflective objects made of hematite, obsidian, or pyrite material stand out within the universe of pre-Hispanic artifacts for their aesthetics, their beauty, and their complexity of production (Blainey 2007; Gallaga 2001, 2009; Healy and Blainey 2011; Pereira 2008; Salinas 1995). Yes, these artifacts were probably also used for vanity purposes in domestic contexts, to see the perfection or imperfections of the onlooker's facial features or to see what cosmetic or jewelry to use. But this was not the only purpose or objective to create and own a mirror. Due to their capacity for projecting an inverse reflection of the spectator's reality (where right becomes left and vice versa), mirrors were used as divinatory or magical portals to communicate between parallel dimensions, worlds, or realities (figure 1.1). With this idea in mind, mirrors were also endowed with the capacity to be a means of contact with the ancestors and more importantly with the gods. It is not hard to imagine complex ceremonial procedures accompanied by chants and dances in secluded locations, perhaps involving fasting and/or the ingestion of psychoactive substances. Such rituals might have been required in order to prepare and train the body and mind to be in contact with the spirits; with the help of the mirror, one presumes that such spirits' advice, guidance, or support was sought out when making important decisions about a course of action to follow. Whether as a ruler, adviser, priest, shaman, or just a *brujo* or *curandero*, the individual or group of individuals who performed these types of actions, envisioned as necessary tasks for the common good of the community, would thereby have acquired great prestige or social position.

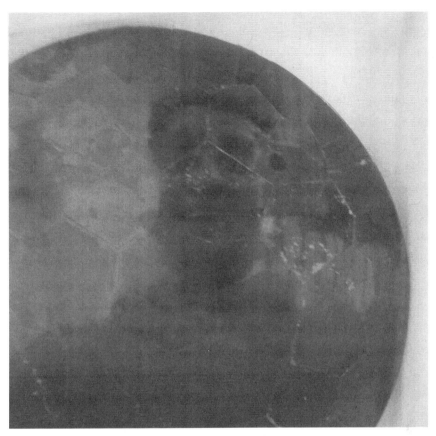

FIGURE 1.1. *Reflection from the Bonampak pyrite mirror (photo by Emiliano Gallaga).*

Although past studies have acknowledged the difficulty of manufacturing these mirrors as well as their importance as objects of prestige and magical-religious worldview, very little research has been carried out concerning how ancient iron-ore mirrors were constructed. Dealing with the issue of mirror production, Emiliano Gallaga (chapter 2, this volume) presents preliminary results of an experimental archaeological project that has the aim of reproducing the operative chain of pyrite mirror manufacture using possible pre-Hispanic tools and techniques. Preliminary results illustrate that this process could take an average of 800–1200 person-hours, representing between 100 and 150 working days for a single person to make an encrusted pyrite mirror. Melgar, Gallaga, and Solis (chapter 3, this volume) also tackle this important question, and present a technological analysis of the manufacturing traces that were applied on different pyrite inlays, using experimental archaeology and scanning electron microscopy (SEM). This methodology allows the authors to identify the lithic tools employed in the production of mirrors with great accuracy and to distinguish different technological styles—fundamental advancements for the study of mirrors, their uses, and the definition of Mesoamerica's artifact assemblage.

In the social realm the possession of iron-ore items would most certainly bestow a high status or social distinction on the owner, not only due to the object's magical-religious connotations, but also for their rarity and cost of manufacturing (Blainey 2007; Gallaga 2001; Pereira 2008; Sugiyama 1992; Taube 1992). In general, items that provide a reflection of an image were not a common thing in ancient times, and yet they were conspicuously present among pre-Hispanic elites. Although pre-Hispanic artisans knew about and used metals, the use of metals was not as vital as that of other materials. Thus, the recognition of mirror craftsmanship is greater if we note the fact that the makers of mirrors almost completely lacked metal tools to fashion the finished mirrors. Due to both their highly symbolic/religious meaning/use and the cost of manufacture, we can infer that mirrors were not a common item to be found on the local markets at the plazas of pre-Hispanic communities. On the contrary, the production of mirrors was most likely restricted and controlled by elites. The craftspeople who made the mirrors would equally enjoy some prestige or recognition not only among the pre-Hispanic elites, but also among fellow artisans as well. As an example, regarding pyrite production at the site of Cancuén, Guatemala, Brigitte Kovacevich (chapter 4, this volume) addresses the techniques and social implications of producing pyrite artifacts. Kovacevich make the case that these objects could have represented high-status goods, ritual paraphernalia, gifts, inalienable possessions, and symbols of individual and collective identities among Cancuén Maya elites. A similar approach is followed by Gazzola, Gómez Chávez, and Calligaro (chapter 5, this volume) for the majestic site of Teotihuacan. The authors describe the archaeological context of thousands of objects, some of them pyrite items, deposited as apparent offerings through the ritual closure of a tunnel under the Feathered Snake Temple, the most important building in the site's Ciudadela Complex. In addition to the lack of prior research on mirror manufacture, other general problems such as lack of archaeological work in various cultural areas of ancient Mexico, lack of information about workshops for mirror production, the incorrect identification of these objects, the looting of sites, and the lack of reporting and publication of archaeological projects, makes for a very poor scholarly record of such materials. Some of these issues are addressed by Gazzola et al. (chapter 5, this volume) with their description of lapidary workshops at La Ventilla, located to the south of the old city. The remains of these workshops enable the study of raw materials, cut waste, stone and bone tools, a few finishing objects, and abrasives for understanding and interpreting the techniques employed in the manufacture of pyrite and hematite mirrors at Teotihuacan.

As a child, I remember a scene from a "western" film I saw in which Apache Indians used mirrors to communicate the arrival of the cavalry in the desert landscape of Arizona. This capacity for reflection, whereby sunlight can be caught or reflected, makes mirrors appear as an evocation of divine or diabolic qualities; in fact, as acknowledged by Lunazzi (chapter 6, this volume), some researchers claim that ancient iron-ore mirrors can set fires if one knows how to use them (see also Ekholm 1972, 1973). One can imagine the effect that the sudden appearance of a fire with the use of a mirror would have among an astonished audience: is the supernatural spirit of the sun trapped in the mirror? Is it the power of the mirror's holder that commands the sun to shine inside the mirror? Although these are not the questions Lunazzi addresses, he does present his experimental results on the reflective capacity of pre-Hispanic mirrors, the real possibility of using mirrors as communication devices, and the renowned ability of these objects to ignite fires. In a somewhat different approach to the concept of lustrous items as solar reflectors, Joseph Mountjoy presents the description of 49 iron-pyrite ornaments. Recovered from his excavations made between 2001 and 2005 in three Middle Pre-Classic period cemeteries in the Mascota valley of Jalisco, Mexico, Mountjoy dates these objects in the range of 1000 to 700 BC, among the oldest such items yet found in Mesoamerica (chapter 7, this volume). Mountjoy contends that these artifacts played a symbolic role in early agricultural societies that were ritually focused on three interrelated factors for survival: sun, water, and fertility, factors that are also symbolized in ornaments of emerald green jadeite and transparent quartz. In chapter 8 (this volume), Achim Lelgemann presents material, technical, and morphological aspects of archaeological mirror remains recovered from an elite burial inside the pyramid of the Citadel patio compound at the site of La Quemada, Zacatecas, dating to the Late or Terminal Classic period (eighth and ninth centuries AD). Lelgemann discusses these mirrors' mortuary-ceremonial contexts, as well as both their functions (as status markers, divinatory devices, lighters) and their socio-ideological dimensions (cosmograms, sun-fire cult, and shamanism) as compared to similar finds in Mesoamerica and the Greater Southwest.

But how did the peoples of the ancient New World actually conceptualize iron-ore artifacts we now call "mirrors"? As presented by Marc Blainey (chapter

9, this volume), it is reasonable to construe these iron-ore mirrors as evidence for shamanistic practices in ancient Maya society. Blainey uses archaeological and iconographic data, as well as ethnographic information from the modern Maya, to illustrate what he calls the "reflective surface complex" in Maya ritual. Similarly, John J. McGraw (chapter 10, this volume) follows Blainey's research path, but with the little twist of focusing on crystals as reflective surfaces that are important to the modern Maya. As we know, crystals have long played a role in Maya ritual. In particular, McGraw demonstrates how divination makes use of crystals to render a series of visual signs that can be interpreted by the diviner as communications from supernatural beings.

Concerning research from areas outside Mesoamerica, Carrie Dennett and Marc Blainey (chapter 11, this volume) address the issue of iron-ore "mirrors" found in Lower Central America, most likely of Maya origin, and how these prestige items arrived at such distant locales. The authors argue for a concept of developing "peer elite" relationships and reciprocity in the form of "gifting," instead of a focus on economic trade factors, which appears to parallel more general sociopolitical and socioeconomic restructuring occurring simultaneously in both areas. Of course, the Maya are not the only people known to use reflective objects as a means of seeing or communicating with other realms, but, unfortunately, there is not much research about the magic/ritual use of reflective surfaces among other Mexican Indian communities. In addressing this gap in the literature, Olivia Kindl's contribution on the ritual use of mirrors among the Huichol Indians of Mexico's West Sierra Madre (chapter 12, this volume) allows the reader to gain a different perspective on the use of these items by a living Indian group outside the Mesoamerican realm. The fact that Kindl had the luxury of speaking with shamans or curanderos who still use mirrors for their ceremonies today, and that she could actually see and participate in those celebrations, provides an intimate perspective full of ethnographic information that can inform the otherwise indirect evidence analyzed by archaeologists. For example, in examining encrypted phrases on pots and stelae, Blainey (chapter 9, this volume) goes to great lengths to identify possible candidates for the Maya glyphs that were in some way associated with mirrors (e.g., T24/T617 "reflective stone" or *ilaj* "was seen"). In a more contemporary mode, Kindl (chapter 12, this volume) obtains similar results from the direct quotes of a present-day Huichol curandero who still uses mirrors for divinatory activities (*xikiri* "things that shine," *nierika* "gift of seeing").

In closing, Karl Taube (chapter 13, this volume) applies his considerable expertise in a critical summary of mirror objects found among ancient and modern Mesoamericans. As Taube makes plain, these objects provide archaeologists and anthropologists with an exceptional opportunity for understanding broader norms of past and present-day Mesoamerican culture, an opportunity that has been overlooked for too long.

MIRRORS AND THE MESOAMERICA CONCEPT

In 1943, a publication shook the minds of all the archaeologists who worked in what at that time was known as "Middle America." That publication was *Mesoamerica: Its Geographical Limits, Ethnic Composition, and Cultural Character* by Paul Kirchhoff (1967), based on a series of investigations undertaken by the International Committee for the Cultural Distribution in America Studies created by the XXVII International Congress of Americanists. Through this delineation of a new region called "Mesoamerica," Kirchhoff's intention was to note what the communities and cultures of a specific area of the American continent share in common and what they do not share (Kirchhoff 1967: 1). Decades later, it is now clear that this work not only achieved its original objective, but it also coined a new term that fills a previous gap in the research areas of Mexico, Central America, and parts of the United States.

The novelty of this proposal was the creation of a term that was not based solely on geographical data, as was common in those days, but on three cultural trait groups: those exclusively for Mesoamerica, those that were present in and outside Mesoamerica, and those that were not present in Mesoamerica. For the first group, 43 traits were considered, such as hieroglyphic writing; use of chinampas (i.e., "floating gardens"); tiered temples; cultivation of maguey, corn, beans, and cacao; and pyrite mirrors. It is interesting to note that from these 43 traits, only 12 were movable artifacts, while the rest are concepts, foods, or architectural structures. Among the diverse array of objects created by pre-Hispanic artisans, it is notable that mirrors (especially iron-ore mirrors) were among the few objects that Kirchhoff selected as archetypes of Mesoamerican culture. I think that this is due to the fact that he considered that mirrors effectively represent the advanced cultural development of ancient Mesoamerican society.

Although Kirchhoff's proposal defines a new cultural region (Mesoamerica), this was not his real intention. He really intended to present a proposal that had to be analyzed, criticized, and/or supplemented by other researchers, preferably archaeologists. However, for the most part that input did not materialize as researchers adopted the term without much hesitation. Indeed, some revisions on Kirchhoff's proposal did appear, such as the critiques by Litvak (1992) and Matos (1994), which focused on the spatiotemporal distribution of the cultural traits Kirchhoff defined and the sources where those traits were obtained. Although it is not the purpose of this volume to tackle the validation of the Mesoamerica concept, it is necessary that we make mention of Kirchhoff's reference to mirrors as illustrating the social complexity of ancient Mesoamerica.

In his publication, Kirchhoff (1967) provided a series of cultural traits that according to him define what Mesoamerica is and what it is not. The critiques leveled by Litvak and Matos are not about the list per se, but more about the origin and the organization of the list. These critical reviewers said that Mesoamerican traits came from different sources, such as ethnography, linguistics, ethnohistory, and archaeology, but not from the material culture context alone (Litvak 1992: 82). Moreover, Matos (1994: 56) stated that there is not a ranking system on Kirchhoff's (1967: 55) trait list to provide a sense of which traits are more Mesoamerican than others. Neither was there an explanation nor a description of what Kirchhoff understood as a cultural trait. Such delimitations could help clarify the geographical range of the trait or the cultural expansion of it. Matos makes an interesting case about this point with the example of the chinampas trait: considering that in the 1940s the chinampas could be found only in the Mesoamerica region, but that later on in the 1990s these agricultural systems were found at Lake Titicaca in Bolivia, does this finding mean that the Lake Titicaca region is part of Mesoamerica? It is understandable that more traits would have to be found in order to make that claim, but the point is that most of Kirchhoff's cultural traits can and are found in other regions and cultures that do not have anything to do with Mesoamerican culture. So, where is borderline of the Mesoamerica region? A ranking of the traits could help, but that is apparently still in the making. Suffice it to say that Kirchhoff's traits refer to a specific pre-Hispanic society that is not described, but is presumed to be a complex one (Litvak 1992; Matos 1994). Yet Mesoamerica is anything but a uniform region, culturally speaking. In Mesoamerica there have always been complex societies living or interacting side by side with less complex communities. This is especially true in the northern areas where interaction and exchange between hunter-gatherer groups was essential for the development of cultural and economic exchange in the region.

A second major critique of Kirchhoff's proposal is the analysis of the spatiotemporal distribution of traits that define Mesoamerica, which is geared toward the time that the pre-Hispanic world came to an end, that is to say the contact period (Kirchhoff 1967: 3). All the traits used by Kirchhoff came from Spanish descriptions and accounts of the pre-Hispanic communities that the

Spanish encountered, as well as ethnographic and some archaeological data, but all of this from the contact period. In Kirchhoff's original proposal there is not an analysis of the cultural development of the Mesoamerican concept through time, that is to say for the Preclassic, Classic, and Postclassic periods. It is important to clarify that this omission or oversight is not imputable to Kirchhoff. As I mentioned before, he made a proposal that had to be built upon and refined by others. In this regard, Litvak (1992) and Matos (1994) make a preliminary analysis of how Mesoamerica should look though time, understanding that Mesoamerica is a cultural and not a geographical area. Just to mention an example, during the Preclassic period Mesoamerica is constrained to the Olmec communities of Tabasco (Litvak 1992) and Guerrero (Matos 1994), and involved in interaction with other soon-to-be Mesoamerican areas. This continues until we reach the Mesoamerica map that we recognize today for the contact period. Litvak clearly summarized this position: "a region identified as Mesoamerican for one phase, could be left out in another" (1992: 89).

I concur with both Litvak and Matos that these critiques do not diminish Kirchhoff's Mesoamerica concept, but rather they serve to strengthen it by providing new elements to see the Mesoamerican area developing through time and space. As a final remark, Litvak concluded his 1992 article, stating that "future work and now non-existent data can modify the concept of Mesoamerica's physical size and shape and even extend it in time, in any direction, without altering the definition" (1992: 102).

For the particular case of this volume's focus on iron-ore mirrors, and the above discussion about the Mesoamerican realm, I will provide a description of what a pyrite mirror represented for ancient Mesoamerican peoples, the elements of which it is composed, and how it is distinguished from other reflective surfaces in other cultural regions, such as that of the Incas. Furthermore, I will present the cultural development of mirrors though the pre-Hispanic periods. This temporal analysis represents ongoing research performed not only by me but by other colleagues as well. As with the case of Kirchhoff's chinampas trait, such ongoing and future research will most likely improve upon the results here presented.

COMPOSITION OF A PYRITE MIRROR

In general, a pyrite-encrusted mirror consisted of four basic elements or characteristics: a base, an adhesive layer, pyrite plaques, and perforations (figure 1.2A).

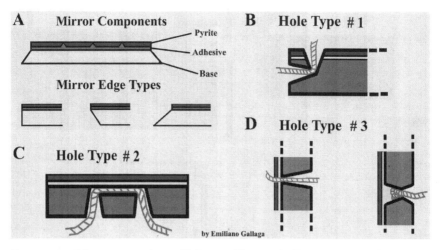

FIGURE 1.2. *Mirror components and hole types (drawing by Emiliano Gallaga).*

BASE

Pyrite mirrors usually consist of pyrite plaques that are adhered to a solid base that is commonly made of stone, like sandstone, mud rock, or slate. There are reports of wood and ceramic bases but, for their lack of preservation, these are not common in the archaeological record of Mesoamerica and the American Southwest/Northwest Mexico (Gladwin et al. 1938; Kelley 1971; Kidder et al. 1946). Often circular or rectangular bases are the norm, but some (very uncommon) triangular bases do exist (Gladwin et al. 1938: plate CIX *e* and *f*). Dimensions of circular bases may range from 7 to 30 cm in diameter with an average of 8–10 mm in thickness. The edges can be perpendicular; beveled inside; and/or beveled outside. The beveled edge can face the front or back of the mirror (see Figure 1.2.A). On some occasions, this area is decorated with painted stucco or with pseudo-cloisonné technique, like those found in Snaketown, Arizona (Gladwin et al. 1938, plate CXI). The backs of mirrors can be decorated with painted stucco and/or direct carving (Blainey 2007; Di Peso 1956, 1974; Ekholm 1945; Furst 1966; Gallaga 2001, 2009; Gladwin et al. 1938; Kelley 1971; Kidder et al. 1946; Smith and Kidder 1951). During the early Postclassic (AD 900–1200), pyrite mirrors were encrusted or framed on a wooden base, which was then decorated with other materials such as jade, turquoise, gold, copper, cotton, or even feathers, like those mirrors found at Chichén Itza, Yucatan, commonly mistaken or described as mosaic disks (Blainey 2007; Coggins 1989; Gallaga 2001, 2009; Taube 1992, Pereira 2008).

Adhesive Layer

Chemical compositional analyses of adhesives have been performed only recently. Most descriptions of how the plaques were attached to their bases are from researchers' guesswork. In general, the description by Kidder et al. (1946) offers the most accepted explanation: the adhesive layer was "a very fine clay, which presumably had been bound and rendered strongly adhesive by mixing it with some organic glue" (Kidder et al. 1946: 126). However, recent conservation efforts regarding a pyrite mirror that was found with turquoise decoration at the center of the Palacio Quemado at Tula, Hidalgo, Mexico, provided one of the first chemical descriptions of an adhesive layer (Magar and Meehan 1995). The researchers mention that "the adhesive used for the turquoise tesserae was composed by a mixture of wax, a [type of] resin, and calcium sulfate" (Magar and Meehan 1995: 7; author's translation). They state further that "an adhesive composed of tar" was employed for the pyrite plaques (Magar and Meehan 2001: 7). In addition, a chemical compositional analysis of adhesive samples from two pyrite mirrors found at the site of Aguateca, Guatemala, showed that lime plaster or stucco was used as an adhesive[1] as well (Keochakian 2001; Takeshi Inomata, personal communication 2004). Keochakian (2001: 11) states that in terms of the adhesive's chemical composition at Copan (Honduras) "a pair of ear flares found in the Subjaguar tomb had jade inlays set into white stucco plaster-like material." Another artifact, found in the Margarita tomb, was interpreted as a possible wooden (?) cup with jade inlays and evidence of white stucco plaster-like "adhesive" (Keochakian 2001: 11).

Some mirrors exhibit an adhesive layer that is yellowish in color. This leads one to suspect that artisans used tree resin as organic glue. This hypothesis has been confirmed by recent analysis made on the pre-Hispanic turquoise-mosaic items mostly from the British Museum (McEwan et al. 2006). Other recent studies on pre-Hispanic mural painting confirm the same, as researchers found that pre-Hispanic artisans used the secretion from orchids (specifically *Cyrtopodium macrobulbun* and *Catasetum maculatum*, known as *ch'it ku'uk* among the Mayas) as organic glue, support, or adhesive for the paintings and in other crafts as well, such as mirror manufacture (Vázquez de Ágredos Pascual 2010: 128). Sometimes the yellowish layer is mistaken for pigment and several mirrors have been misidentified as pigment mortars (Kelley 1971). Depending on the mirror, the adhesive layer can be 1–3 mm thick.

Pyrite Plaques or Tesserae

Pyrite is a yellowish mineral made of iron and sulfur or iron sulfide (FeS_2). The common shape of pyrite is cubic but it can appear in other polygonal shapes (Lagomarsino 2008: 121). Due to the material's characteristics, the shaping of iron pyrite pieces is a time-consuming and very skill-demanding task—the greatest of the entire mirror-production process (Gallaga 2009) (figure 1.3). As Kidder et al. (1946) state:

> Pyrite, with a hardness of 6.5 and with no natural cleavage planes to facilitate subdivision of the crystals, could not have been other than most difficult to work. Yet every plaque was mounted with dozens or scores of plates cut precisely the same thickness and shaped to fit exactly. The polygons seldom had less than four and some possessed as many as nine sides, each so beveled that only the very edge came into contact with that of its neighbor (Kidder et al. 1946: 131).

Due to its instability, water action and oxygen can transform the pyrite into other minerals like iron oxide (limonite and siderite). Because of that, it is difficult to recover this material in good condition within archaeological contexts (Zamora 2002a, 2002b: 695).

The number of pyrite plaques used for a single mirror varies from specimen to specimen and is thought to range from one to as many as 40 or 50 pieces (Furst 1966; Taube 1992; Turner 1992). The dimensions of the plaques may vary from one to four square inches, with an average thickness of between 2 and 4 mm. Also, dimensions probably depended on the availability and type of raw material, as well as on the intended size and design of the mirror. Some mirrors have been reportedly made with a single piece of pyrite. Apparently, there is a source where the vein of pyrite is attached to a layer of sandstone, so a block was removed, giving it the shape and polish of a mirror without using adhesive (Mata Amado 2003, Mike Jacobs personal communication 2001).

The face where the pyrite plaques were applied could cover the entire front surface of the stone base, but also could leave free a surface of 1 or 2 cm wide at the edge of the mirror. Sometimes this edge could be beveled and on some occasions they were decorated, as discussed above. Pyrite plaques become hematite if exposed to fire and on many occasions they might have been mistaken as pigment (Gladwin et al. 1938; Smith and Kidder 1951, Woodbury and Trik 1953).

Holes or Perforations

An important characteristic of pyrite-encrusted mirrors is the hole that is made to wear or suspend the object in some way. Generally, researchers

FIGURE 1.3. *Sample of pyrite plaques or tesserae from a mirror found at tomb 10 of building 21, from the site of Tenam Puente, Chiapas, dated to the Late Classic period (Martínez del Campo Lanz 2010: 76–77) (photo by Emiliano Gallaga).*

describe two locations for the holes: at the edge and at the center. A combination of both is common and may have held some decorative or other functions (Kidder et al. 1946; Smith and Kidder 1951; Taube 1992). Currently, there is no standardized typology for classifying these perforations; however, researchers commonly comment on the presence/absence of holes. For instance, in order to provide a more effective description for the pyrite mirrors of Snaketown, a typology for holes and perforations was made, based upon their manufacture and existing hole descriptions from other sites and research projects. Three types of holes constitute this typology (Gallaga 2001, 2009):

Type 1 consists of a pair of perpendicular interconnected holes made at the edge of the mirror. Generally, there are two pairs of holes at the opposite edge, but it is possible to have one pair of holes per side. Also, depending on the function of the mirror one pair of holes may be found at the upper portion of the mirror (figure 1.2.B).

Type 2 corresponds to "a pair of holes near the center of the backing, connected by a shallow groove which allowed the cord to pass beneath and be hidden by the encrustation" (Smith and Kidder 1951: 48). In this type, a cord has to be strung through the perforations before the pyrite encrustation process begins (figure 1.2.C).

Type 3 is commonly known as suspension holes. These are perforations that go straight through. Those can be at the center, at opposite edges, or both. In single-hole mirrors, the end of each cord must be secured by a knot or toggle (figure 1.2.D) (Smith and Kidder 1951).

None of these perforation types are exclusive, and combinations of types in one single mirror are common. Combinations may be the result of function and/or decoration/adornment of the mirrors themselves. This last feature could be the result of the different forms of use, or combinations of functions and/or the type of decoration found on the mirror. For example, the perforations of types 1 or 2 could be associated with the mirrors used in the occipital portion of the lower human back that are known as *tezcacuitlapilli* (Gallaga 2001) and generally associated with warriors, members of the elite, high ranking priests, and ambassadors (see Blainey 2007; Kidder, et al. 1946; Sugiyama 1992; Taube 1992).

MESOAMERICAN MIRRORS THROUGH TIME

Although studies concerning these materials are relatively new, we have already begun to establish a historical development of these devices, which can be coupled to the standard Mesoamerican periods (Ekholm 1973; Gallaga 2001, 2009; Pereira 2008):

MIDDLE PRECLASSIC PERIOD (1200–400 BC)

The first mirrors recorded in archaeological context are those located in the Olmec region, particularly at the site of La Venta (Heizer and Gullberg 1981; Pires-Ferreira and Evans 1978). Such mirrors are characterized by being manufactured with metallic minerals (magnetite, hematite, and ilmenite) in one piece with a finely polished concave surface, and in some cases with holes, quite possibly for use in hanging. Generally it is considered that this type of concave mirror was used for the diffraction of sunlight and to light a fire (Ekholm 1973). Regardless of how these objects came to form part of the Olmec magical-religious structure, their appearance and use eventually spread throughout the rest of Mesoamerica and beyond (Blainey 2007; Clark and Hansen 2001; Grove 1977; Pereira 2008).

EARLY CLASSIC PERIOD (AD 150/200–600)

In this phase, the mirrors' manufacture underwent a radical transformation: they begin to have a flat surface, rather than concave, and they are not made in one piece, but instead feature a stone base, upon which polygonal pyrite plaques are arranged in a mosaic. This change represents a technological breakthrough and innovation, since the manufacture of mirrors allowed more aesthetic freedom to play with the designs of the mosaic tiles of pyrite and in some cases decorations on the posterior base of stone. From the archaeological evidence collected so far, it is inferred that most of the mirrors for this phase were made in Teotihuacan or made to imitate this style (Ekholm 1973; Pereira 2008; Taube 1992). However, we cannot rule out that other major manufacturing centers existed, such as in the Oaxaca region (Pires-Ferreira 1975; Pires-Ferreira and Evans 1978; Mohar 1997). This interest in pyrite by the Teotihuacan people is also seen in other reflective materials such as mica, used to make mirrors or adornments.

In a recent discovery in a royal tomb at Chiapa de Corzo, Chiapas, archaeologists found two square mirrors, each with a flat surface composed of several thick plates of pyrite fitted with a thick stucco layer over a decomposed organic base (probably wood) and approximately 2,700 years old (700–500 BC). This find indicates that mirrors made with several pieces of plaques were already being constructed somewhere in Mesoamerica much earlier than originally thought. However, these pyrite plaques from Chiapa de Corzo are rectangular in shape and very thick, much different from the thin polygonal plaques conventionally used on the mirrors identified for the Early Classic period. Also, this is the earliest report for this type of pyrite plaque on a mirror, so the working hypothesis is that we have encountered a transitional mirror specimen, in between the Preclassic- and Classic-period styles (Gallaga and Lowe 2012). In other words, somebody somewhere started a new way to make mirrors with pyrite plates instead of single pieces of magnetite, hematite, or ilmenite (Olmec style), and probably that is why the pyrite plates are somewhat less elaborate on this Chiapa de Corzo mirror.

EARLY POSTCLASSIC PERIOD (AD 900–1200)

After several centuries without significant changes, mirrors underwent another radical transformation: the mirror of the previous phase was incorporated into a larger base, usually of wood, which was decorated with intricate baroque mosaics of different materials like turquoise, obsidian, shell, copper, and gold. The most notable example of this phase is the disk located at

Chichen Itza (Blainey 2007; Gallaga 2001; Pereira 2008). In parallel, we find in the Tarascan region a mimicry of these mirrors, but instead of wooden bases with mosaics, Tarascan wooden bases are covered with a sheet of copper or bronze, on which the tiles were incorporated (Pereira 2008; Di Peso 1974). Some examples of this variation have been found in northern Mexico, particularly at the site of Paquimé, Chihuahua (Di Peso 1974). In general, the encounter with these innovations intensifies the idea of the magical-religious messages encoded in these mirrors, with more surface area to decorate and enhance the aesthetic value and status of the object.

LATE POSTCLASSIC PERIOD (AD 1200–1521)

In the process of incorporation of metal, there was a gradual replacement of pyrite mosaic mirrors, in exchange for gold and copper discs with intricate turquoise tiles (Pereira 2008). In this regard, it is interesting to note the absence of pyrite mirrors alongside the presence of gold disks with turquoise mosaics, in the relationship of objects rendered to the Triple Alliance (Sepúlveda y Herrera 2003). However, pyrite mirrors are present in the *Codex Kingsborough*, as a tribute from the Oaxacan region (Mohar 1997). Similarly, Sahagún (1989) mentions the presence of mirrors in Aztec markets, which denotes a more popular use, probably for more domestic vanity purposes, with the implication that these objects had acquired a less exclusive status by the time of European contact.

CLOSING REMARKS

As I have illustrated here and as will be found throughout this volume, iron-ore mirrors are among the most sophisticated items produced by pre-Hispanic artisans or craftspeople. These mirrors were made in a time before a glass was coated with a tin-mercury amalgam process,[2] which had to be imported later from Europe. If one is to follow Kirchhoff (1967), iron-ore "mirrors" are among a specific list of artifacts that characterize or even define the Mesoamerican region. Even though researchers long ago recognized the complexity and symbolism implied in the use of these intriguing items, little research exists regarding their social significance, function, or the precise steps in the manufacturing process that produced them. Lately, the importance of pre-Hispanic mirrors as prestige and/or magical-ritualistic items has been coming under increased discussion, as have the more technological aspects of mirror manufacture (Blainey 2007; Gallaga 2001, 2009; Healy and Blainey 2011; Pereira 2008; Salinas 1995). The advancement of the current research makes the present volume a timely venture, as it provides a more comprehensive analysis of these shiny objects, integrating different aspects of mirror manufacture, use, and symbolism, as well as conducting a reexamination of the question as to what extent such "mirrors" define and/or characterize the Mesoamerican region (figure 1.4).

In order to arrange the great variety of contributions, the volume is divided into three main sections. The first section (chapters 1–5) focuses on the production aspects of mirrors, with chapters ranging from experimental archaeology projects to discussions of workshops in archaeological contexts in the Maya, Central Mexico, and Northwest Mexico regions. The second section (chapters 6–9 and chapter 11) concentrates on the question of the use and meaning of mirrors during pre-Hispanic times. Special attention is given to the use of such items as both sacred and luxury artifacts. The last section (chapter 10 and chapters 12–13) centers on the use of mirrors leading into modern times by contemporary indigenous communities, with emphasis on examining and stressing the relationship between ethnographic reality and archaeological interpretation.

Owing to the multidimensional importance of mirrors in ancient and present-day Mesoamerican societies, any scholarly study of these objects requires an interdisciplinary approach. Hence, although this volume commences to analyze iron-ore mirrors according to their technical aspects, the chapters proceed from experimental results to the social domains of archaeology, anthropology, and iconography. In this way, one witnesses how these mirror objects reflect the social scientific study of indigenous Mesoamericans more broadly. As scholars continue to elucidate the significance of these objects for the human groups who made and used them, we encounter foreign worldviews and ways of life that are just as complexly human as our own.

NOTES

1. At least for one of the mirrors. The analysis for the second mirror was inconclusive, suggesting it might have been organic resin (Takeshi Inomata personal communication 2004; Keochakian 2001).

2. The use of silver-mercury amalgams to make mirrors started as far back as AD 500 in China. However, it was not until the fourteenth century that the process to coat a glass with a tin-mercury amalgam was perfected by European manufacturers (Pendergrast 2003: 14, 31).

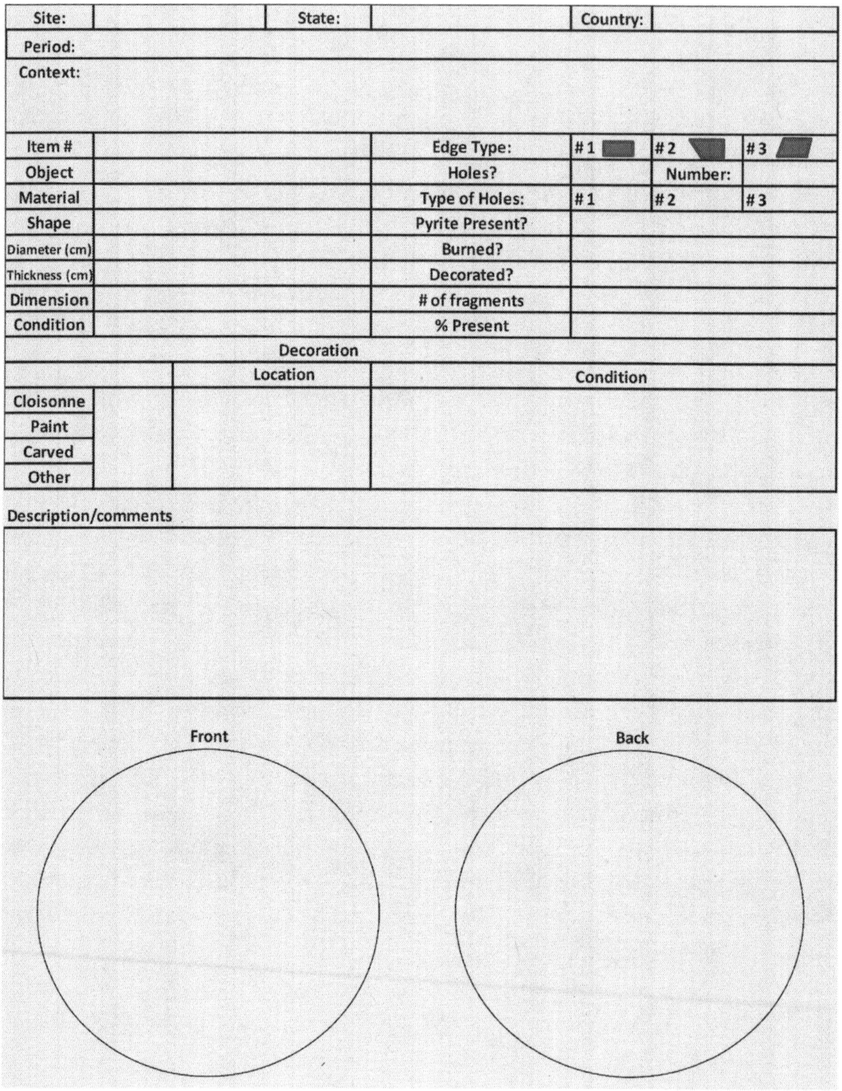

Site:		State:		Country:	
Period:					
Context:					

Item #		Edge Type:	#1	#2	#3
Object		Holes?		Number:	
Material		Type of Holes:	#1	#2	#3
Shape		Pyrite Present?			
Diameter (cm)		Burned?			
Thickness (cm)		Decorated?			
Dimension		# of fragments			
Condition		% Present			

Decoration				
	Location		Condition	
Cloisonne				
Paint				
Carved				
Other				

Description/comments

Front   Back

FIGURE 1.4. *Proposed registration sheet for pyrite mirrors (made by Emiliano Gallaga).*

REFERENCES

Albenda, Pauline. 1985. "Mirrors in the Ancient Near East." *Notes in the History of Art* 4 (2–3): 2–9.

Baboula, Evanthia. 2000. *Bronze Age Mirrors: A Mediterranean Commodity in the Aegean*. Nicosia, Cyprus.

Beasley, J. D. 1949. "The World of the Etruscan Mirror." *Journal of Hellenic Studies* 69: 1–17. http://dx.doi.org/10.2307/629458.

Blainey, Marc. 2007. "Surfaces and Beyond: The Political, Ideological, and Economic Significance of Ancient Maya Iron-Ore Mirrors." MA thesis, Department of Anthropology, Trent University. Peterborough, Ontario: The European Association of Mayanists; http://www.wayeb.org/download/theses/blainey_2007.pdf.

Bulling, A. 1960. *The Decoration of Mirrors of the Han Period*. Ascona, Switzerland: Artibus Asiae Publishers.

Cameron, Fiona. 1979. *Greek Bronze Hand-Mirrors in South Italy*. BAR International Series 58. Oxford: British Archaeological Reports.

Cammann, Schuyler. 1949. "Chinese Mirrors and Chinese Civilization." *Archaeology* (2): 114–120.

Clark, John E., and Richard D. Hansen. 2001. "The Architecture of Early Kingship: Comparative Perspectives on the Origins of the Maya Royal Court." In *Royal Courts of the Ancient Maya, Volume 2: Data and Case Studies*, edited by Takeshi Inomata and Stephen D. Houston, 1–45. Boulder, CO: Westview Press.

Coggins, Clemency, and Orrin C Shane. 1989. *El Cenote de los sacrificios: tesoros mayas extraidos del cenote sagrado de Chichen Itzá*. Mexico: Fondo de Cultura Economica.

Di Peso, Charles C. 1974. *Casas Grandes: A Fallen Trading Center of the Grand Chichimeca*. Vol. 1. Dragoon, AZ: Amerind Foundation.

Di Peso, Charles C. 1956. *The Upper Pima of San Cayetana del Tumacacari: An Archaeological Reconstruction of the Ootal of Pimeria Alta*. Dragoon, AZ: Amerind Foundation Publication 7.

Ekholm, Gordon F. 1945. "A Pyrite Mirror from Queretaro, México." Carnegie Institute of Washington Notes for Middle American Archaeology and Ethnology, no. 53. Washington, DC: Carnegie Institute of Washington.

Ekholm, Gordon F. 1972. "The Archaeological Significance of Mirrors in the New World." Paper presented at the Atti del XL Congresso Internazionale degli Americanisti, Roma-Genova.

Ekholm, Gordon F. 1973. *Abstractos de la 40vo Congreso Internacional de Americanistas, Rome-Genoa*. vol. 1., 133–135. Genova, Italy: Casa Editrice Tilgher.

Furst, Peter T. 1966. "Shaft Tombs, Shell Trumpets and Shamanism: A Culture-Historical Approach to Problems in West Mexican Archaeology." PhD Dissertation, Department of Anthropology University of California, Los Angeles.

Gallaga M., Emiliano. 2001. "Descripción y análisis de los espejos de pirita del sitio de Snaketown, AZ, EU y su relación con Mesoamérica." Paper presented at the XXVI Mesa Redonda de Antropología, Zacatecas, Zacatecas, México.

Gallaga M., Emiliano. 2009. "La manufactura de los espejos de pirita: Una experimentación." Paper presented at the 33rd Congreso Internacional de Americanistas, Mexico City, July.

Gallaga, Emiliano, and Lynnet Lowe. 2012. "Cuentas, Ollas y Espejos: El Recinto Funerario de uno de los Señores de Yoquí (Chiapa de Corzo), Chiapas." Paper presented at the XVIII Congreso Internacional de Antropología Iberoamericana, San Luis Potosí, March 29–31.

Gladwin, H. S., E. W. Haury, E. B. Sayles, and N. Gladwin. (Original work published 1938) 1965. *Excavation at Snaketown: Material Culture.* Medallion Papers no. 25. 2nd ed. Tucson, AZ: University of Arizona Press.

Grove, David C. 1977. "Olmec Archaeology: A Half Century of Research and Its Accomplishment." *Journal of World Prehistory* (11): 51–101.

Healy, Paul F., and Marc G. Blainey. 2011. "Ancient Maya Mosaic Mirrors: Function, Symbolism, and Meaning." *Ancient Mesoamerica* 22 (2): 229–244. http://dx.doi.org/10.1017/S0956536111000241.

Heizer, Robert F., and Jonas E. Gullberg. 1981. "Concave Mirrors from the Site of La Venta, Tabasco: Their Occurrence, Mineralogy, Optical Description, and Function." In *The Olmec and Their Neighbors*, ed. Elizabeth P. Benson, 109–116. Washington, DC: Dumbarton Oaks Research Library and Collections.

Kelley, J. Charles. 1971. "Archaeology of the Northern Frontier: Zacatecas and Durango." In *Handbook of Middle American Indians: Archaeology of Northern Mesoamerica*, Part two, edited by Robert Wauchope, vol. 11: 768–801. Austin: University of Texas Press.

Keochakian, Sylvia. 2001. "Technical Study: Adhesive Material Used on Pyrite Mirrors from Aguateca." Unpublished technical report from the Smithsonian Center for Materials Research and Education, Suitland, MD.

Kidder, Alfred V., Jesse D. Jennings, and Edwin M. Shook. 1946. *Excavation at Kaminaljuyu, Guatemala.* Publication 561. Washington, DC: Carnegie Institution of Washington.

Kirchhoff, Paul. 1967. *Mesoamerica: Sus límites geográficos, composicion étnica y caracteres culturales.* Mexico: suplemento de la revista Tlatoani.

Lagomarsino, James. 2008. *A Pocket Guide to Rocks and Minerals.* Bath: Parragon.

Lilyquist, Christine. 1979. *Ancient Egyptian Mirrors*. Munich: Deutscher Kunstverlag.

Litvak, Jaime. 1992. "En torno al problema de la definición de mesoamerica." In *Una definición de Mesoamerica*, ed. Jaime Litvak, 74–105. Mexico: UNAM.

Magar, Valerie, and Patricia Meehan. 1995. "Investigación para la interpretación y la conservación de un disco de mosaico de turquesa." Licenciatura thesis, Departament of Restauración de Bienes Muebles, ENR-Churubusco, México.

Magar, Valerie, and Patricia Meehan. 2001. "Consideration on the Conservation and Investigation of the Turquoise Mosaic Disk from the Palacio Quemado, Tula, México." Paper presented at the 66th SAA Annual Reunion, New Orleans, LA.

Martínez del Campo Lanz, Sofía. 2010. *Rostros de la divinidad: Los mosaicos mayas de piedra verde*. México: INAH.

Mata Amado, Guillermo. 2003. "Espejo de pirita y pizarra de Amatitlán." In *XVI Simposio de Investigaciones Arqueológicas en Guatemala, 2002*, edited by J. P. Laporte, B. Arroyo, H. Escobedo, and H. Mejía, 831–839. Guatemala: Museo Nacional de Arqueología y Etnología.

Matos, Eduardo. 1994. "Mesoamerica." In *Historia Antigua de Mexico, Vol. 1*, edited by Linda Manzanilla and Leonardo López Luján, 49–74. Mexico: INAH, UNAM, Ed. Porrua.

McEwan, Colin, Andrew Middleton, Caroline Cartwright, and Rebecca Stacey. 2006. *Turquoise Mosaics from Mexico*. London: The British Museum Press.

Mohar, Luz María. 1997. *Manos Artesanas del México Antiguo*. México: SEP, CONACYT.

Pendergrast, Mark. 2003. *Historia de los Espejos*. España: Ediciones B.

Pereira, Gregory. 2008. "La Materia de las Visiones: Consideraciones acerca de los espejos de pirita prehispánicos." *Diario de Campo*, mayo/junio (48): 123–136.

Pires-Ferreira, Jane W. 1975 "Formative Mesoamerican Exchange Networks with Special Reference to the Valley of Oaxaca." In *Prehistory and Human Ecology of the Valley of Oaxaca*, vol. 3, edited by Kent. V. Flannery. Museum of Anthropology Memoirs No. 7. Ann Arbor, MI: University of Michigan.

Pires-Ferreira, Jane W., and Billy Joe Evans. 1978. "Mössbauer Spectral Analysis of Olmec Iron Ore Mirrors: New Evidence of Formative Period Exchange Networks." In *Cultural Continuity in Mesoamerica*, ed. David L. Browman, 101–154. The Hague, Netherlands: Mouton. http://dx.doi.org/10.1515/9783110807776.101.

Sahagún, Fray Bernardino de. 1989. *Historia General de la Cosas de la Nueva España*. México: Alianza Editorial Mexicana, Fondo de Cultura Económica.

Salinas, Flores. 1995. *Tecnología y Diseño en el México Prehispánico*. Facultad de Arquitectura, Centro de Investigaciones de Diseño Industrial. México: UNAM.

Sepúlveda y Herrera, Maria Teresa. 2003. "La Matrícula de Tributos." *Arqueología Mexicana* (14): 1–85.

Smith, A. Ledyard, and Alfred V. Kidder. 1951. *Excavations at Nebaj, Guatemala.* Publication 594. Washington, DC: Carnegie Institution of Washington.

Sugiyama, Saburo. 1992. "Warfare, and Human Sacrifice at the Ciudadela: An Iconography Study of the Feathered Serpent Representation." In *Art, Ideology, and the City of Teotihuacan*, ed. Janet Berlo, 205–230. Washington, DC: Dumbarton Oaks Research Library and Collection.

Taube, Karl A. 1992. "The Iconography of Mirrors at Teotihuacan." In *Art, Ideology, and the City of Teotihuacan*, ed. Janet Berlo, 169–204. Washington, DC: Dumbarton Oaks Research Library and Collection.

Turner, Margaret H. 1992. "Style in Lapidary Technology: Identifying the Teotihuacan Lapidary Industry." In *Art, Ideology, and the City of Teotihuacan*, ed. Janet Berlo, 89–112. Washington, DC: Dumbarton Oaks Research Library and Collection.

Tolkien, John Ronald Reuel. 1991. *The Lord of the Rings.* London: BCA.

Vázquez de Ágredos Pascual, María Luisa. 2010. *La Pintura Mural Maya: Materiales y Técnicas Artísticas.* Mexico: UNAM.

Woodbury, Richard B., and Aubrey S. Trik. 1953. *The Ruins of Zaculeu, Guatemala.* vol. 1–2. Richmond, VA: William Byrd Press.

Zamora, Fabian Marcelo. 2002a. "La industria de la pirita en el sitio de Aguateca durante el periodo Clásico Tardío." Licenciatura thesis, Department of Archaeology, Universidad del Valle de Guatemala, Guatemala.

Zamora, Fabian Marcelo. 2002b. "La industria de la pirita en el sitio Clásico Tardío de Aguateca." In *XV Simposio de Investigaciones Arqueológicas en Guatemala 2001*, ed. J. P. Laporte, H. Escobedo, and B. Arroyo, 695–708. Guatemala: Museo Nacional de Arqueología y Etnología.

2

The research objective that concerns us here is to determine the mechanisms of manufacture of pyrite mirrors and establish the implications of these artifacts in the social, political, and religious realm of ancient Mesoamerica. The questions are many: what are the origins of these mirrors? How were raw materials obtained? By what techniques were mirrors developed? How many person-hours were required for the manufacture of mirrors? What tools were used to produce them? How can we identify the workshops where mirrors were made? What was their distribution and use? Who manufactured them?

To begin to answer some of these questions, I performed a project of experimental archaeology in which I tried to recreate the working or operational chain of a pyrite mirror, an endeavor that has not been attempted before in modern times. Although the project has not yet achieved the ultimate goal, which is the total recreation of a mirror, this chapter summarizes the achievements so far in the reconstruction of the manufacture sequence and presents preliminary results of experiments for manufacturing this type of artifact.

PRE-HISPANIC CRAFT PRODUCTION

Study of manufacturing and/or production of objects are not limited to economic factors. Research also covers the culture of the people who produce and consume materials, because it is culture that frames the demands

How to Make a Pyrite Mirror

An Experimental Archaeology Project

EMILIANO GALLAGA M.

DOI: 10.5876/9781607324089.c002

and uses of products. Behind every one of the objects made in ancient Mexico, there existed an accumulation of knowledge of raw materials, tastes and fashions, transportation, and trade networks. Some of these objects also evoke a magical-religious complex of cultural structures and practical solutions for everyday life that characterize the society that created them. Although the study and analysis of material remains is a common subject in archaeology, only in recent decades has there emerged intensified research into craft production, where experimental archaeology has played a major role, much like the inclusion of innovative technologies and specialists from other fields of knowledge (Costin 1991, 2001; Manzanilla 2006; Velázquez 2007; Melgar Tísoc 2008).

Although some objects reach the needs of local markets and common demands, others are intended for use by a specific, high-ranking social stratum and therefore require more exotic materials and greater craft specialization to produce, and might circulate in markets or circuits outside the region. Thus, at some point in the pre-Hispanic period, certain elite individuals, groups, and/or communities started to live partially or totally dissociated from subsistence activities, which allowed some people to devote their time exclusively to the manufacture of particular objects for which they were paid, or as part of previously established tax payments (Costin 1991, 2001; Velázquez 2007). Hence, as summarized by Velázquez (2007: 17), it "is only possible to speak of expertise when there are more consumers than producers." This social differentiation or specialization may have been a direct result of increases in the complexity of social stratification that led to the emergence of specialists in each and every one of the social sectors, including handicraft production. In addition, the growing demand for luxury and exotic objects that denote the high status of the wearer/owner had to be satisfied. This process most likely started with the Olmecs during the Middle Preclassic period (1200–400 BC).

In the investigation of craft production, Cathy Costin (1991) identifies four general parameters to define the degree of production of a certain item, craft unit, or artisan. In general, these parameters can provide a good idea about where the production of pyrite mirrors was situated among pre-Hispanic communities in Mesoamerica. These parameters are:

Context: the complexity of production and distribution. At one extreme are independent artisans, whose production, especially popular consumer devices, is subject to market fluctuations. Located at the opposite end are craftspeople dependent on or sponsored by the elites and the ruling class for their manufacture of luxury goods.

Concentration: the level of concentration of the means of production and production itself. On one side we have the dispersion of production units or workshops with high variability in the objects produced, and low production volumes, while at the opposite end we have a concentration of production units, with standardization and mass production of objects. Here the quality and type of items completed in the production units should also be taken into account.

Scale: the dimensions of the production unit and the number of members who compose it. This concept is much more difficult to establish because few workshops are located in the archaeological context and the evidence found—for example, waste material or garbage—is not always straightforward. In addition, there have been very few cases in which those special contexts have been described and published.

Intensity: the time spent on the development of an object from the collection of the raw material until completion of the work to make it. These scales of person-hours (time production) could provide a good idea of the manufacturing costs of an item and its possible cost/value in the market.

In addition to these four general patterns, in a subsequent study Costin (2001) identifies three other important factors to consider for the study of pre-Hispanic production: (1) *objects*: what assets are used in the manufacture of objects and what prestige or restrictions (if any) are entailed in their use? (2) *principles and mechanisms of distribution*: how are the goods produced being transported to consumers, what are the distances and types of transaction (if any), and are those operations voluntary? and (3) *consumers*: who is using the final product (which in some way determines the use and value of the artifact)?

To locate the intensity or levels of the patterns determined by Costin, it is also important to know the *technology* used to manufacture the object. Velázquez (2007: 20) defines this term as "the set of social and material elements with which man modifies his environment [including] tools and the products that are made from them, and the knowledge, behaviors, attitudes, and meanings as well, which are shared by groups of people [and] transmitted from one generation to another" (translation by author). Thus, knowledge of the processes used in the preparation of a particular object allows us to ascertain with more security the levels of production and specialization, as well as to identify the tools used, the waste material, and the material contexts for unit productions or workshops. The establishment of these contexts can in turn determine or imply patterns of specialization and standardization, with which one can speak of levels of skill and efficiency of the artisans (Costin 1991, 2001; Melgar Tísoc 2008; Schiffer 1992; Velázquez 2007).

Such analyses of the objects allow us to begin to understand the processes that must have been performed for their manufacture and the possible tools used. However, for many cases of pre-Hispanic production, we do not know the steps of the manufacturing process or operational chain, the means of production, or the production units. In order to establish such manufacturing processes, it is necessary to experimentally reproduce the object and obtain results similar to archaeological materials in question. This is where experimental archaeology enters.

EXPERIMENTAL ARCHAEOLOGY

Experimental archaeology works from the assumption that every single item manufactured in the past was regulated by cultural patterns of production that provide particular characteristics to the object in question. So, the reproduction of such manufacturing processes, as well as the tools and techniques used, must obtain similar results to those objects made in the past with similar production traces or patterns (Ascher 1961; Schiffer et al.1994; Velázquez 2007). In general, experimental archaeology projects follow two lines of research: (1) those that try to establish the process that affected an item as exhibited at the time of discovery (such as conditions like exposure to fire, chemical, or other taphonomic processes), and (2) those studies that attempt, successfully or not, to reproduce the manufacturing steps, or production process, of an item in order to establish production times, signatures of manufacture, material waste, or labor experience such as efficiency, variability, and standardization (Ascher 1961; Schiffer et al. 1994; Tringham 1978; Velázquez 2007). The present chapter focuses on the second type of research.

In some cases it is possible to apply the knowledge generated by ethnoarchaeologies, especially when there is a relative continuity of production process by a living community, or when such production was recorded in some way in the past. The use of these analogies (discontinuous or continuous) is assumed to be registering production processes of living communities that may be similar to those found in ancient cultures that have since disappeared and whose only record is the material context (Binford 1991; Gándara Vázquez 1990; Maldonado 2005).

In the case of iron-ore mirrors, it appears that their production stopped immediately after the introduction of European mirrors into New World markets. I have no record of any community that has preserved or continued to manufacture these objects after this initial period of contact. For this reason, there is no direct ethnoarchaeological material that could be referenced in this project, although I nevertheless consult ethnographic information regarding lapidary work (Langenscheidt 2006; Maldonado 2005; Mirambell 1968; Pastrana 1998; Velásquez 2008). Similarly, there is some data available from ethnohistorical sources, such as Sahagún and Durán, who note the names that were given to some of the precious stones, and certain equipment used in the production of lapidary items, such as Tototepec and Quetzaltepec sands (Durán 1867:425) and powders from Huaxtepec flints (Oaxtepec, Morelos) to carve precious stones, or the use of bitumen made with bat guano for polishing obsidian mirrors (Sahagún 1989). Despite this wealth of information, few studies have focused on known archaeological techniques and tools with which the societies of ancient Mexico worked precious and semiprecious stones. Therefore, experimental archaeology proffers data that would otherwise be difficult, if not impossible, to retrieve. Following the work of researchers who have previously conducted experimental archaeology projects, research must begin with a phase of exploratory experiences that allows the researcher to stipulate variables, or the problems to be considered in a later phase of controlled experimentation. In the first phase, one registers, describes, and classifies the tools and materials used, the different manufacturing steps followed, and the time required for production. The results form the basis for establishing specific goals achievable in the next phase.

The controlled-experiments phase will consider manufacturing variables to establish patterns of production, working to identify and compare manufacturing fingerprints with archaeological material. On the latter point, it is important to emphasize the use and application of optical devices to compare the samples macro- and microscopically. If you use similar materials, tools, and techniques to those of pre-Hispanic artisans, manufacturing traces must be similar in both samples from current and pre-Hispanic times (GibajaBao 1993; Schiffer et al. 1994; Velázquez 2007, 2008).

IRON-ORE MIRRORS

Among the vast universe of objects made by pre-Hispanic artisans, mirrors are definitely one of the objects that demand more attention, in part for their aesthetics, but more for the skills needed to manufacture them (especially if we take into account that metal tools were not worked profusely in

Mesoamerica). Depending on the period, the most common materials used in the manufacture of these mirrors were hematite (Preclassic period), pyrite (Classic period), and obsidian (Postclassic period) (Ekholm 1973; Gallaga 2001, 2009; Pereira 2008). In the particular case of the hematite (Fe_2O_3; the name derives from the Greek for "blood"), this is a mineral found in rhombohedral crystals, and represents the most common source of the world's iron and steel. It is also frequently the basis of ocher pigments employed in the creation of red paint (Blainey 2007: 167–168; Hurlbut and Sharp 1998: 177–178). Hematite occurs with two distinctive appearances: at times it is compact and possesses a black or grayish black color and brilliant metallic luster, but much of the hematite worked as an ore is loose and earthy in form and red in color (Blainey 2007: 167–168). Although, hematite has a metallic luster and provides a good reflection, it is more unstable than pyrite. We know that in China, around AD 500, they used silver-mercury amalgams to make mirrors, but there is not enough evidence to suggest the use of such amalgams in mirrors in Mesoamerica, although mercury was known there. Glistening, shiny, liquid mercury, or "quicksilver," has been recorded from at least six different ancient Maya sites, in amounts ranging from 90 to over 600 g, and associated with ritual caches or elite burials containing jade, cinnabar, pearl, shell, bone, and ceramic remains (Austin 1994: Table 2). Though at least one of these caches dates to the Late Preclassic period, all other occurrences are from the Classic period (AD 250–900) (Healy and Blainey 2011: 241, footnote 8).

In this chapter, the research focuses on pyrite-mosaic mirrors, which in general consisted of four basic elements: base, adhesive layer, pyrite plaques, and perforations or suspension holes (see chapter 1 and figure 1.1) .Researchers concur that the person-hours needed to produce at least one of these artifacts must be high, as these mirrors were some of the most complex items made by pre-Hispanic artisans (Blainey 2007, Di Peso 1974; Furst 1966; Gallaga 2001; Kidder et al. 1946; Pereira 2008; Smith and Kidder 1951; Woodward 1941; Zamora 2002a, 2002b). The comments of Smith and Kidder (1951: 44) on the mirrors excavated from the site of Nebaj, Guatemala, embody the best observations of these materials:

> These objects were marvels of painstaking craftsmanship. Much time must have been required to produce their remarkably even-surfaced stone backings, in most cases so almost perfectly circular as to suggest the use of some compass-like device. But this was as nothing compared to the work of cutting and grinding to exactly equal thickness, and of beveling for accurate edge-to-edge fit, the

many polygonal plates of refractory pyrite crystals with which the face of each was incrusted. One hesitates to guess at the amount of labor that went into the making of even one such plaque. (Smith and Kidder 1951: 44)

For their material characteristics and the presumably magical-religious values they had, mirrors were considered by Kirchhoff (1943) as one of the characteristic elements for defining the Mesoamerican region.

In his investigation of these objects in the Maya area, Blainey (2007) determined that about 63% of iron-ore mirrors from the Maya region are related to power or religious contexts, whereas other contexts are associated with fills or surface collections in plazas or in caves.[1] Similar behavior is seen in the many pyrite mirrors (of clear Mesoamerican manufacture) located in archaeological contexts in a completely different region from the Maya; in Northern Mexico and the Southwestern United States these objects were being used in similar ways as in Mesoamerica, that is, in a very close relationship with the local spheres of power and magical-religious structures (Gallaga 2001, 2009). With respect to their distribution, pyrite mirrors must have enjoyed wide demand as luxury, magical-religious, or decorative status-markers, since they are located in regions stretching from the Southwestern United States and Northern Mexico through Mesoamerica and Central America to Peru (Blainey 2007; Di Peso 1974; Furst 1966; Gallaga 2001, 2009, Gladwin et al. 1938; Kidder et al. 1946; Lothrop 1937; Pereira 2008; Smith and Kidder 1951; Woodward 1941; Zamora 2002a, 2002b).

Mirrors had many more features in ancient times than the narrow vanity function they serve today. In recent decades, a consensus has been reached that pyrite mirrors had been used mainly for ritual divination and magical-civic activities, such as communicating with the ancestors, serving as portals to alternate realities, to start fires or reflect light beams, as part of clothing, or as social symbols or prestige objects used in ceremonies (Blainey 2007; Ekholm 1945; Gallaga 2001, 2009; Pereira 2008; Taube 1992; Zamora 2002a; see also chapter 11, present volume).

Clearly it required a huge investment in person-hours to prepare one of these objects, since it demanded very specific materials and skilled artisans, most likely under the supervision of elite members. It is not hard to imagine that the acquisition and use of these items should have been restricted to highly select social sectors. The representations of persons carrying or having one of these mirrors are limited to military roles, rulers, ambassadors, *pochtecas*, and other elites, or sometimes slaves or servants who hold them for their masters.

EXPERIMENTAL PHASE

This experimental archaeology project focuses on the category of research that reproduces the steps of manufacturing a device, which can measure production times, traces of use, waste material, or work experiences such as efficiency, variability, or the standardization of production (Ascher 1961; Schiffer et al. 1994; Velázquez 2007). Due to the almost total lack of previous work on experimentally manufacturing pyrite mirrors, with the exception of some attempts to reproduce a pyrite plaque (Ekholm 1973; Zamora 2002a), I decided to follow the basic procedures of experimental archaeology for exploratory experiences aimed at recording, describing, and classifying the tools and materials, the different manufacturing steps, and the time required in production. Based on the results, I can evaluate and design a second phase of the experiment with procedures and defined targets (Ascher 1961; GibajaBao 1993; Schiffer et al. 1994; Velázquez 2007).

The first step was to identify the raw material used for the production of the object, possible tools used, and possible steps or process for its manufacture. This procedure follows the steps determined by the Experimental Archaeology Project of Marine Shell Material at the Templo Mayor, directed by archaeologist Adrián Velázquez (2007), who has established a set of manufacture patterns, the tools used, and the traces left by those tools for shell objects.

It is important to note that there is a lack of information about the production units or work spaces where mirrors were made in pre-Hispanic times. In the archaeological literature of the areas where these items have been found, only five sites have been identified as possible mirror manufacture units: (1) San José Mogote, Oaxaca, where they made hematite mirrors in the Preclassic period (Pires-Ferreira 1975; Pires-Ferreira and Evans 1978); (2) Teotihuacan, Mexico (Turner 1992; Widmer 1991); (3) Aguateca, Guatemala (Zamora 2002a, 2002b); and (4) Cancuén, Guatemala (Barrientos, personal communication 2008; see also chapter 4, this volume) for pyrite-mirror production in the Classic period. The fifth site, Pacbitun, Belize, has been identified only as a stone-base workshop (Zamora 2002a; Fialko 2000) (figure 2.1). Unfortunately, information about the tools and the contexts of production is limited and often superficial. In addition to these archaeological contexts, historical sources were also consulted, such as the *Historia General de las Cosas de la Nueva España* (Sahagún 1989) and the *Códice Florentino* (López Luján 1991) and sources about communities of mirror artisans, such as the *Matrícula de Tributos* (Sepúlveda y Herrera 2003) and the *Códice Kingsborough* (Mohar 1997; Valle 1993).

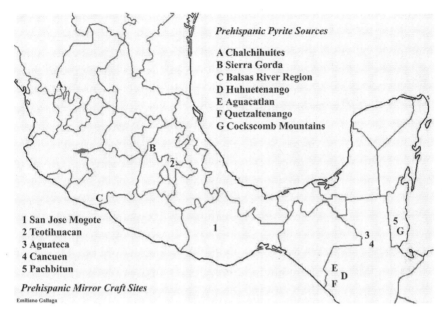

Prehispanic Pyrite Sources

A Chalchihuites
B Sierra Gorda
C Balsas River Region
D Huhuetenango
E Aguacatlan
F Quetzaltenango
G Cockscomb Mountains

1 San Jose Mogote
2 Teotihuacan
3 Aguateca
4 Cancuen
5 Pachbitun

Prehispanic Mirror Craft Sites

Emiliano Gallaga

FIGURE 2.1. *Location of the known pre-Hispanic pyrite sources and mirror-craft production sites (map by Emiliano Gallaga).*

MANUFACTURING PROCESS

The first step was to re-create the stone base. From the information collected, it was established that the most common materials used were slate and sandstone (Blainey 2007; Gallaga 2001; Pereira 2008; Zamora 2002a). Several pieces of this material were collected from local sources found on the cuts of the highway Tuxtla Gutierrez–Tecpatán (figure 2.2A). As for abrading tools, a basalt metate was bought at the local market in Tuxtla Gutiérrez (Figure 2.2B) and sand, rich in quartz, was collected from Puerto Arista, Tonalá, as well as river stones (as hammers) (Figure 2.2C and D).

Once all the raw materials and tools were collected, the process began. From the sandstone pieces collected, one was selected that did not have imperfections or fractures. With the aid of percussion with the stone hammers this sandstone piece was worked into a desirable size. I decided to make a stone base of nine cm in diameter and eight mm thick, which is an average size for the encrusted pyrite mirrors found in archaeological contexts (Blainey 2007; Gallaga 2001; Zamora 2002a). At this point, three person-hours had been spent, not taking into account the time to acquire the materials. The next step was to abrade the piece with sand and water over the metate until it reached

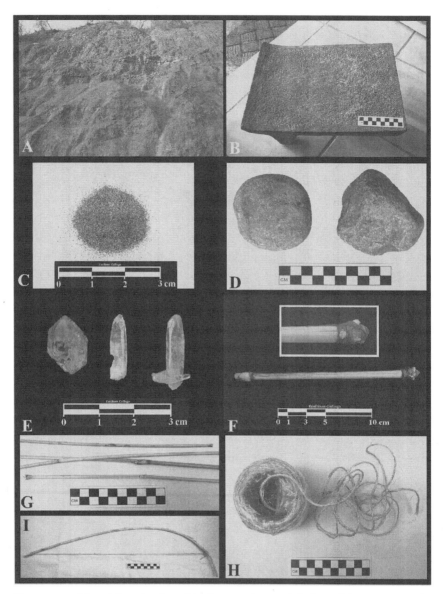

FIGURE 2.2. *Materials and tools used in the manufacture of the base of the mirror:*
(a) Tecpatán sandstone source, (b) basalt metate, (c) sand from Puerto Arista, Tonalá,
(d) choppers, (e) crystal quartz from Zacatecas, (f) cane chisel tipped with a crystal quartz,
(g) canes, (h) rope, and (i) arched bow made with cane (photos by Emiliano Gallaga).

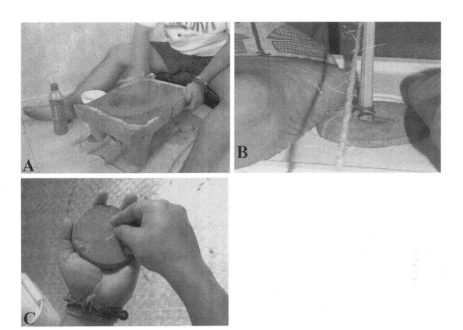

FIGURE 2.3. *Manual manufacturing process of the stone base: (a) carved process of the sandstone piece over a basalt metate with sand and water, (b) drilling holes with a chisel tipped with crystal quartz by pulling a cane bow, and (c) making the slot with crystal quartz (photos by Emiliano Gallaga).*

the desired size, by which time a total of 51 person-hours had been spent (figure 2.3A). Later, the piece was beveled at one cm; this took eight additional hours using the same tools and procedures, bringing the process to a total of 59 person-hours.

At this point, the stone base was almost complete, except for the perforations. From Gallaga's typology (Gallaga 2009; and see chapter 1, present volume), I decided to drill a pair of type 2 holes, placed at the center of the piece and working from the back of it. The holes would then be connected in the front by a canal to be covered by the pyrite plaques (this canal allows for a string to be laced between the holes without disrupting the mosaic surface). For this process, different sizes of Zacatecas quartz crystals were acquired at the La Lagunilla market in Mexico City (figure 2.2E). These crystals were placed at the tip of Carrizo sticks, and later, to give traction to them, a bow was used (figure 2.2.F–I) and the perforations were made (figure 2.3B). Once the holes were finished, the crystals were used to make the interconnection canal between them (figure 2.3C). The perforation process took a little more than 28 hours to complete. In total, the manufacture of a sandstone base with its perforations took 87 person-hours, a little more than 10 days of work (considering eight working hours per day) (figures 2.4A–D). Recent information mentions that the slate bases for mirrors from Teotihuacan show manufacture traces of being worn away with andesite and cut and perforated with chert (Emiliano Melgar, personal communication 2010). Future research will test this recently discovered drill process.

The next step, and the most difficult, was the manufacture of a pyrite plaque. First, at the La Lagunilla market several types of pyrite from Zacatecas and from the center of Mexico were acquired: radial, cubic, and amorphous (figure 2.4E). Once I acquired the raw material, the first real problem arose: which process or manufacture technique should I use to cut the pyrite? As mentioned before, there has yet to be any experimental verification in this regard, only some personal observations or attempts here and there. Zamora (2002a: 22) states that he carried out some tests to make pyrite flakes by direct percussion, obtaining some workable pieces, but he did not proceed to work the pieces. Sahagún (1989: 3:74) mentions that the pre-Hispanic stone artisans cut the fine stones with sand as an abrasive with a hard metal. But later on he mentions that for very hard stones such as "blood chert, being extremely hard, cannot be cut with emery, only broken into pieces with a stone. You take the good parts, the ones that can be polished ... and abrade with water over a hard stone" (making reference to obsidian).

With this information, I inferred that a hard material such as pyrite was worked with percussion techniques. At that time, a number of comments and suggestions pertaining to how to work this material were made by several researchers, but especially from archaeologists Adrián Velázquez and Emiliano Melgar from the Shell Experimental Archaeological Project and the Lapidary from Templo Mayor: Style and Technological Traditions Project. Further advice was received from Gregory Pereira, Marc Blainey, and Roberto Velásquez, who all agreed that percussion would be a good option to start with. At this point of the experimentation, I started to work a pyrite core with a stone hammer through direct percussion techniques, methodology that was considered at the time to be the more reliable. The results were as expected, with several irregular pieces without flat surface due to the particular characteristics of the pyrite. One of those pieces was chosen to be worked by abrading with sand and water over a basalt metate.

Although the process is not finished, the pieces resemble many pre-Hispanic pyrite plaques found archaeologically (figure 2.4F–H). A total

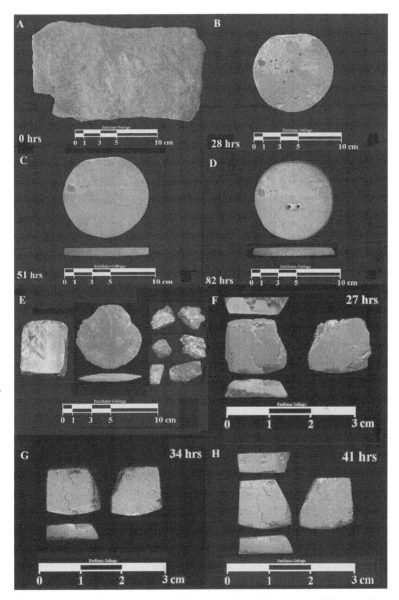

Figure 2.4. *Photographic record of the manufacturing process of the piece of sandstone base and the tile of pyrite: (a) unworked piece, (b) piece with 28 hours of work, (c) with 51 hours of work, (d) final piece with 82 hours of work, (e) different types of pyrite, (f) pyrite fragment with 27 hours of work, (g) with 34 hours of work, and (h) with 41 hours of work (photos by Emiliano Gallaga).*

of 41 person-hours have been invested so far in a single plaque and as can be observed, the methodology used seems to work, but I am conscious that other alternative techniques can be tested in future studies. Not long after performing this initial experiment, eight pre-Hispanic pyrite-plaque samples from a mirror found at the site of Tenam Puente (courtesy of archaeologist Lalo Jacinto) were submitted to Emiliano Melgar for preliminary analysis of manufacturing traces (see figure 1.3, this volume). This analysis used scanning electron microscopy (SEM; Jeol JSM–6460-LV) equipment from the Subdirección de Laboratorios y Apoyo Académico of INAH.

Results from the archeological samples showed diffuse flattened bands of 100 μm width on a rugged texture (figure 2.5a), which are similar to those experimental traces obtained by using basalt metates with sand as an abrasive, and the brightening with leather (figure 2.5b), while the cut marks showed straight edges with a rugged texture, crossed by fine lines between 0.5 and 2 μm width (figure 2.5c), similar to those obtained by experimental cutting with obsidian flakes or blades by Melgar's experimental project (figure 2.5d). At the time, this was the first analysis of pyrite material from a pre-Hispanic context that illustrated such results. Subsequently, further analysis of different samples of pyrite from Teotihuacan, Templo Mayor, and Chiapa de Corzo were made. The results illustrated that the tesserae were cut by flint or obsidian flakes (see chapter 3, present volume). The SEM information corroborates that the pyrite plaques were abraded with sand over basalt, as in the experiments. However, the information about cutting the pyrite with obsidian flakes was surprisingly new and was not tested at the time of this project.

Although the objective to make a complete encrusted pyrite mirror has not yet been achieved, the preliminary results can shed some light on the manufacturing process.

MATERIALS

The production of these items would have been somewhat restricted by the use and functions of social norms and ritual, but also by the skill and time required for their manufacture. Pyrite mirrors would be neither cheap nor easily acquired. The artisan or artisans dedicated to manufacturing this type of object could have been concentrated in communities or in a region near the raw material sources, such as San José Mogote, Oaxaca, in the Preclassic or in communities that had complex trade networks that allowed them to acquire the raw materials, such as Teotihuacan in Mexico and Aguateca, Cancuén, or Pacbitun in Guatemala (Fialko 2000; Mata Amado 2003; Pires-Ferreira

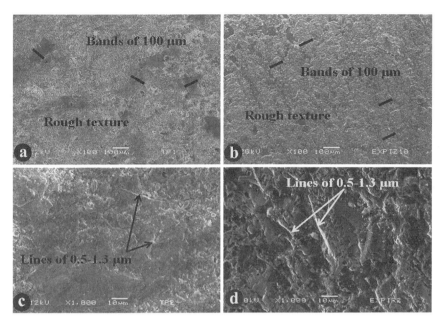

FIGURE 2.5. *Images of scanning electron microscopy (SEM) in which it is possible to observe: (a) surface detail from the pyrite tile from Tenam Puente, (b) surface detail of the pyrite tile carved with basalt and brightened with leather obtained experimentally, (c) cut detail from the pyrite tile from Tenam Puente, and (d) detail of cut made with obsidian blades obtained experimentally (photos by Emiliano Melgar).*

1975; Pires-Ferreira and Evans 1978; Turner 1992; Widmer 1991; Zamora 2002a, 2002b; Barrientos, personal communication 2008).

In the particular case of the stone bases, sandstone and slate are a very common material in Mesoamerica that would not represent a problem to acquire and use. However, different scenarios are presented for the pyrite. Sahagún (1989: 3: 333) mentions that the raw material used for the manufacture of mirrors was extracted from mines, but he never mentions any names or locations. In the literature only 16 pre-Hispanic sources or mines where pyrite material was extracted have been identified: (1) Chalchihuites, Zacatecas; (2) the Sierra Gorda de Queretaro; (3) the Rio Balsas region, Guerrero; (4) Huehuetenango, (5) Aguacatlan, and (6) Quetzaltenango in Guatemala; (7) the Cockscomb Mountains, Belize; (8) the Guatemalan Massive deposit at Chiquimula; (9) the Guatemalan Residual deposit at Izabal and Zacapa; (10) the Chixoy Polochic Vault Zone in Central Guatemala; (11) the Quetzal Mine near Copan; (12) the

Mochito Mine near the Lake Yojoa at Santa Barbara, Honduras; (13) Agalteca near the Honduran Montes de Comayagua; (14) the El Salvadorian deposits; (15) the Northern Nicaraguan deposits; and (16) the Coastal-Shore deposits along the Pacific Coast of Guatemala (Takeshi Inomata, personal communication 2004; Blainey 2007; Kelley 1971; Langenscheidt 1988; Maldonado 1980; Thompson 1939; Weigand 1995, 1997; Zamora 2002a), However, it is likely that these were not the only pre-Hispanic sources used (see figure 2.1). Clearly, while more research on pyrite mines or sources is needed, it seems fair to assume that few places produce the raw material necessary to make pyrite mirrors, providing an additional cost and inherent luxury to the artifact.

STONE BASES

Based on the experimental project, it is fair to say that the work and the necessary time to make the stone base were not especially heavy nor did the task require special skill. Or at least, this does not represent more work than any other pre-Hispanic lapidary work. The process is tedious but can be performed by nonspecialized craftspeople and supervised by a chief artisan. However, more skills or experience would be necessary to make the perforations or the suspension holes. In this process, it took me a total of 87 hours to make a sandstone mirror base, a little more than 10 days of work at eight hours a day by a single person.

A different story is involved in the decoration of the internal face or the beveled areas of the mirrors. The internal face is the one opposite to the pyrite mosaic and the one that is going to be attached to the body of the holder, if the pyrite mirror face is the one the public observers can see. The decoration of this face would be made by a chief artisan, probably one under direct supervision of elites or by special members of the noble classes who not only make the decoration but also performed special rituals for the objects. If we think that these objects are personal items, the decoration could be an individual signature or identification of a social group or elite personage, such as those mirrors from Calakmul, Campeche (Blainey 2007; Kerr 1993), or the pseudo-cloisonné decoration on the beveled edges of the mirrors at Snaketown, Arizona (Gladwin et al. 1938).

Little discussion has been addressed to the function or purpose for these elaborate decorations that, in most cases, would be facing the chest of the wearer. If the patchwork of pyrite is the front of the mirror, facing potential admirers, why would so much work be expended on carving the back of the mirror? Most likely, those decorations must have been codes exclusively for the "eyes of the holder," something very similar to the internal inscriptions of wedding rings, for instance, or the elaborate representations of the Aztec goddess Tlaltecutli, lady of the earth, on the bases of sculptures to be appreciated only by the goddess herself.

Pyrite Plaques

It has been speculated, and with good reason, that the most difficult phase of the mirror manufacture process is the creation of the pyrite plaques. Although in the future it is necessary to try other processes to cut the pyrite, such as with obsidian blades, chert dust, sand, water, and the use of a cord, the percussion technique discussed herein seems to work just fine. However, it is a costly method, losing a lot of the raw material in the process (Sahagún 1989). If percussion was the process used, it can be inferred that more skill and experience is necessary than I presently possess. In the case of cutting the pyrite with obsidian flakes, preliminary results indicate less waste of raw material, but with far greater time required to produce a single plaque (see chapter 3, this volume). Although the main objective of this project has not been achieved to the fullest, the fact that I have succeeded in the complete manufacture of one pyrite plaque provides preliminary yet crucial information about the possible parameters of time and production techniques for this complicated pre-Hispanic item.

At this point, more than 40 person-hours have been invested on the manufacture of one single pyrite plaque, 4 mm thick with a surface of 1 cm square. This plaque is not finished,[2] but it already resembles those found in archaeological contexts. By itself, 40 person-hours do not sound like much, but if we think that on average a mirror used between 20 and 30 pyrite plaques of similar dimensions, the amount of time needed is enormous: an average of 800–1200 person-hours, representing between 100 to 150 working days for a single person to make an encrusted pyrite mirror, not counting broken attempts and decoration. I make the emphasis on a single person, but of course more than one person could have been involved in the process, a concerted effort that would drastically reduce the time of manufacture. It is also necessary to recognize that a more experienced craftsperson should be able to do it in less time. Even so, my experiment of making a single plaque shows us the time-consuming nature of this aspect of mirror manufacture, while also suggesting the considerable skills and focus that would be necessary. With this in mind, and considering the expertly crafted specimens found archaeologically, we can presume that the pre-Hispanic artisans who worked the pyrite were full-time specialists.

A characteristic of the pyrite plaques that always arises in previous publications is their geometric shape (Blainey 2007; Ekholm 1945; Gallaga 2001; Pereira 2008; Taube 1992; Zamora 2002a). Usually a plaque has between four and nine side-edges that fit perfectly with each other. In a practical way, one could easily assemble the mosaic plaques as both square and polygonal. If we follow the principle of using as little material as possible, polygonal plaques are a better optimization of resources. However, if we take into account the type of raw materials and cutting techniques used, the square plaques are much easier to work and to attach to give shape to a mirror regardless of the geometric form of the base. In spite of this, the general trend of pre-Hispanic crafters was to manufacture polygonal plaques on circular bases, which would have been much more laborious to make.

Probably the answer to our question about the ultimate purpose of these objects is not practical, but ceremonial. To show a mosaic with polygonal plaques that present a clear frame on the edge, this might have represented a somewhat abstract shape of a turtle shell, an animal that in ancient times has been associated mythologically with the earth or as a portal that connects with parallel dimensions and/or magical worlds (Blainey 2007; Miller and Martin 2004; Pereira 2008; Schele and Miller 1986). Considering that the pyrite mirrors are one of the cultural elements that define Mesoamerica (Kirchhoff 1943), it would not be unreasonable to suggest a relationship between the mirrors, the mother earth, and their representatives in it. In other words, one can envision the Mesoamerican mirror as a magical representation of a turtle shell, which in turn is a portal between the worlds contained in the mother earth and the underworld.

PRODUCTION UNITS

The experimentation process to re-create an encrusted pyrite mirror is still underway, but enough data has been accumulated thus far that I can venture some preliminary conclusions. At this point, a stone base of 9 cm diameter by 8 mm thickness and one pyrite plaque have been completed. If something between 20 and 30 pyrite plaques are needed to complete one mirror, using the same techniques, a conservative estimation gives us something between 900 and 1300 person-hours or 110–160 days of work for a single artisan to make a single mirror. Due to this estimated amount of time, the accessibility of the raw material, and the necessary skill needed for its manufacture, it is fair to infer that the artisan(s) who worked these items were full-time specialists. Blainey (2007: 107) mentions that "while we cannot rule out the possibility that the mirrors occasionally passed to members of the lower classes, it can be asserted with confidence that the artistic and technical skill required to make the most elaborate mirrors means they were probably restricted to nobles who had the power to commission such objects from royal artisans." In this regard, Sahagún (1989: 65) describes the artisan guild according to their specialization: *tolteca* (abraders who work with stone), the *tecuitlahuaque* (who work with precious stones, gold, and silver), the *tlatecque* (stone cutters), and the *chalchiuhtlatecque* (who finish the precious-stone items). Inside each guild several levels existed according to skill and experience. In addition, Sahagún mentions that this type of job was inherited from father to son, indicating specialization in the manufacture of metal and lapidary items, especially pyrite material, as well as the organization and specialization of the manufacturing process that could be maintained generation by generation. Following this ethnohistorical data, it is not hard to conceive of a few households or production units specializing in the manufacture of these prestige objects, and this is probably the reason why very few production units have been recorded or identified in archaeological contexts.

It was mentioned earlier that only a few sites have been identified as possible workshops: San José Mogote and Teotihuacan, Mexico; Pacbitun, Belize; and Aguateca and Cancuén, Guatemala (Barrientos, personal communication 2008; Fialko 2000; Mata Amado 2003; Pires-Ferreira 1975; Pires-Ferreira and Evans 1978; Turner 1992; Widmer 1991; Zamora 2002a, 2002b), but these sites cannot be the only ones and more research is needed in this regard. In addition, even the limited descriptions that do exist of the material, tools, and spaces of the workshops could shed some light on the manufacturing process for mirrors, such as the research presented in this volume in chapters 4 and 5.

In terms of the parameters for artisan production established by Costin (1991), the following conclusions can be reached with regards to the archaeological data and the preliminary results of this experimental archaeological project:

Context. The production and distribution of the encrusted pyrite mirrors was extremely complex. As is mentioned above, the acquisition of the pyrite raw material could be difficult and expensive due to the limited number of sources available and the difficulty and cost of transporting this raw material to the workshop areas. The artisans dedicated to the manufacture of these items were most likely full-time specialists, dependent on or supported by elites, similar to those described by Sahagún (1989). In a similar way, the distribution of the workshop products (in this case the pyrite mirrors) was restricted to elite members only.

Concentration. The means of production and the actual production are concentrated in a very few locales, with an apparent standardization of the manufacturing process, and with the production of high-quality objects not intended for popular markets.

Scale. Until today, only four possible workshop units had been identified, with very scattered descriptions of them regarding possible work force, tools used, refuse material, and final products. However, it is likely that more than one artisan was involved in the manufacture of a mirror, such that some made the stone base, others the pyrite plaques, and some were in charge of the final decoration of the compound object. If the process of making a pyrite plaque was percussion, it is expected that there would be a great amount of refuse material nearby; but if the cutting of the pyrite was with obsidian flakes the optimization of the raw material is greater, resulting in less pyrite refuse but greater obsidian refuse (figure 2.6). Regardless of the pyrite plaques' manufacturing process, the production cost and time necessary to make a mirror are quite high.

Intensity. Provisional achievements of this project allow me to propose that a single artisan would need between 900 and 1300 person-hours, or between 110 and 160 work days, to make a standard encrusted pyrite mirror. As in any other labor input approximation, these figures can change depending on the skill and experience of the lapidary artisan, on the raw material, and on the manufacturing process used. These estimates are therefore provisional, general, and preliminary, and pending future experiments, such figures should be referenced with caution.

FINAL THOUGHTS

This chapter illustrates only the tip of an iceberg for the manufacture of this special item, the pyrite-encrusted mirror. Until today, the evidence illustrated that the most likely areas of mirror manufacturing in Mesoamerica are found in three regions: Oaxaca, the Maya region, and/or the Central Mexico region. It is important to consider that the location of production, as well as the trade routes and the sources to obtain the pyrite, all could have changed across time. The preliminary results on person-hours needed to make a single mosaic mirror indicate that a high amount of time and energy was required in mirror manufacture. Most likely, the production of these objects was exclusive and

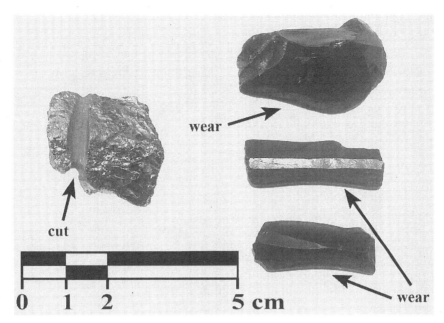

FIGURE 2.6. *Experimental cutting of a pyrite piece with obsidian blades: 60 hours of work.*

restricted to full-time artisans organized into specialized guilds. The tools and process of manufacture patterns are not yet definitively identified and more studies need to be conducted regarding this vital area of research. Several lines of experimentation still need to be performed, such as cutting pyrite with obsidian blades to test its effectiveness, person-hours, and efficiency. However, I have presented this initial work of experimental archaeology as a groundbreaking baseline to initiate future research into the manufacturing process of Mesoamerican iron-ore mirrors.

Acknowledgments: I thank the archaeologists Adrián Velázquez and Emiliano Melgar for their support, comments, and the opportunity to analyze some samples at the Subdirección de Laboratorios y Apoyo Académico del INAH. Likewise, thanks go out to the archaeologist Lalo Jacinto, who provided samples of pyrite tesserae from Tenam Puente site; to Ofelia Murrieta for his comments about his lapidary experience and for providing some raw materials; and to Marc Blainey, Gregory Pereira, and Roberto Velázquez for their much-appreciated comments. I also thank Dr. Gillian Newell and Xavi Gael for their comments and support. Finally, to Sistema Nacional de Investigadores for its financial support of this research.

NOTES

1. From a sample of 175 contexts where mirrors have been located, 64 (36.6%) are from burials, 46 (26.3%) from caches, 19 (10.9%) from fills, 6 (3.4%) found in surface collections, and 40 (22.9%) are from various contexts such as caves, mountains, plazas, *chultunes*, and general excavations (Blainey 2007: 113).

2. I think that the final process for the manufacture of the plaques occurred when all the mosaic pieces had been made and then the whole mirror was polished to make an even surface.

REFERENCES

Ascher, Robert. 1961. "Experimental Archaeology." *American Anthropologist* 63 (4): 793–816. http://dx.doi.org/10.1525/aa.1961.63.4.02a00070.

Austin, Patricia A. 1994. "Mercury and the Ancient Maya." Master's thesis, Department of Anthropology, Trent University, Peterborough, ON.

Binford, Lewis R. 1991. *Bones, Ancient Men and Modern Myths.* London: Academic Press.

Blainey, Marc. 2007. "Surfaces and Beyond: The Political, Ideological, and Economic Significance of Ancient Maya Iron-Ore Mirrors." MA thesis, Department of Anthropology, Trent University. Peterborough, Ontario: The European Association of Mayanists; http://www.wayeb.org/download/theses/blainey_2007 .pdf.

Costin, Cathy Lynne. 1991. "Craft Specialization: Issues in Refining, Documenting and Explaining the Organization of Production." In *Archaeological Method and Theory*, vol. 3. ed. Michael Schiffer, 1–56. Tucson: University of Arizona Press.

Costin, Cathy Lynne. 2001. "Craft Production Systems." In *Archaeology at the Millennium*, ed. Gary M. Feinman and T. Douglas Price, 273–327. New York: Sourcebook, Kluwer Academic/Plenum Publishers. http://dx.doi.org/10.1007 /978-0-387-72611-3_8.

Di Peso, Charles C. 1974. *Casas Grandes: A Fallen Trading Center of the Grand Chichimeca*. vol. 1. Dragoon, AZ: Amerind Foundation.

Durán, Fray Diego de. 1867. *Historia de las Indias de Nueva España y islas de Tierra Firme*. México: Imprenta de J.M. Andrade y F. Escalante.

Ekholm, Gordon F. 1945. "A Pyrite Mirror from Queretaro, México." In *Carnegie Institute of Washington, Notes for Middle American Archaeology and Ethnology*, 2 (53): 178–181. Washington, DC: Carnegie Institute of Washington.

Ekholm, Gordon F. 1973. *Abstractos de la 40vo Congreso Internacional de Americanistas, Rome-Genoa*. vol. 1., 133–135. Genova, Italy: Casa Editrice Tilgher.

Fialko, Wilma. 2000. "El espejo del entierro 49: morfología y texto glífico." In *El Sitio Maya de Topoxté: Investigaciones en una isla del Lago Yaxhá, Peten, Guatemala*, edited by Wolgang W. Wursters, 144–149. Materialien zur Allgemeinen und Vergleichenden Archaologie, Vol. 57. Bonn: Verlag Philipp von Zabern, Mainz am Rhein.

Furst, Peter T. 1966. *Shaft Tombs, Shell Trumpets and Shamanism: A Culture-Historical Approach to Problems in West Mexican Archaeology*. PhD diss., Department of Anthropology, University of California, Berkeley, CA.

Gallaga M., Emiliano. 2001. "Descripción y análisis de los espejos de pirita del sitio de Snaketown, AZ, EU y su relación con Mesoamérica." Paper presented at the XXVI Mesa Redonda de Antropología, Zacatecas, Zacatecas, México.

Gallaga M., Emiliano. 2009. "La manufactura de los espejos de pirita: Una experimentación." Paper presented at the 33rd Congreso Internacional de Americanistas in Mexico City, July.

Gándara Vázquez, Manuel. 1990. "La analogía etnográfica como heurística: Lógica muestral, dominio etnográfico e historicidad." In *Etnoarqueología, Primer Congreso Bosh Gimpera*, edited by Yoko Sugiura and Maricarmen Serra, 43–82. IIA, México: UNAM.

GibajaBao, J. F. 1993. "El comó y el porqué de la experimentación en análisis funcional." *Revista de Arqueología* 148:10–15.

Gladwin, H. S., E. W. Haury, E. B. Sayles, and N. Gladwin. (Original work published 1938) 1965. *Excavation at Snaketown: Material Culture*. Medallion Papers no. 25. 2nd ed. Tucson: University of Arizona Press.

Healy, Paul F., and Marc G. Blainey. 2011. "Ancient Maya Mosaic Mirrors: Function, Symbolism, and Meaning." *Ancient Mesoamerica* 22 (02): 229–244. http://dx.doi.org/10.1017/S0956536111000241.

Hurlbut, C. S., Jr., and W. E Sharp. 1998. *Dana's Minerals and How to Study Them*. 4th ed. New York: John Wiley & Sons.

Kelley, J. Charles. 1971. "Archaeology of the Northern Frontier: Zacatecas and Durango." In *Handbook of Middle American Indians: Archaeology of Northern Mesoamerica*, part two, edited by Robert Wauchope, vol. 11: 768–801. Austin: University of Texas Press.

Kerr, Justin. 1993. "A Precolumbian Portfolio." Accessed June 16, 2009. http://research.mayavase.com/kerrportfolio.html.

Kidder, Alfred V., Jesse D. Jennings, and Edwin M. Shook. 1946. *Excavation at Kaminaljuyu, Guatemala*. Publication 561. Washington, DC: Carnegie Institution of Washington.

Kirchhoff, Paul. 1943. "Mesoamérica; sus limites geográficos, composición étnica y caracteres culturales." *Acta Americana* (1): 92–107.

Langenscheidt, Adolphus. 1988. *Historia Mínima de la minería en la Sierra Gorda*. México: Rolston-Bain, Windsor.

Langenscheidt, Adolphus. 2006. "Los abrasivos en Mesoamérica." *Arqueología Mexicana* 14 (80): 55–60.

López Luján, Leonardo. 1991. "Peces y moluscos en el libro undécimo del *Códice Florentino*." In *La Fauna del Templo Mayor*, edited by Oscar Polanco, 213–263. México: INAH, GV editores.

Lothrop, Samuel Kirkland. 1937. *Coclé, an archaeological study of central Panama*. Memoirs of the Peabody Museum of Archaeology and Ethnology 7. Cambridge, MA: Harvard University.

Maldonado, Blanca. 2005. "Análisis tecnológico de la metalurgia prehispánica de Michoacán: Etnoarqueologia y experimentación." In *Etnoarqueologia: El contexto dinámico de la cultura material a través del tiempo*, edited by Eduardo William, 215–236. Zamora: El Colegio de Michoacán.

Maldonado, Ruben. 1980. *Ofrendas asociadas a entierros del Infiernillo en el Balsas*. México: Colección Científica, SEP-INAH.

Manzanilla, Linda. 2006. "La producción artesanal en Mesoamérica." *Arqueología Mexicana* 14 (80): 28–35.

Mata Amado, Guillermo. 2003. "Espejo de pirita y pizarra de Amatitlán." In *XVI Simposio de Investigaciones Arqueológicas en Guatemala, 2002*, edited by J. P. Laporte, B. Arroyo, H. Escobedo, and H. Mejía, 831–39. Guatemala: Museo Nacional de Arqueología y Etnología.

Melgar Tísoc, Emiliano R. 2008. *La explotación de recursos marinos-litorales en Oxtankah*. México: INAH.

Miller, Mary Ellen, and Simon Martin. 2004. *Courtly Art of the Ancient Maya*. Fine Arts Museum of San Francisco. New York: Thames and Hudson.

Mirambell, Lorena. 1968. *Técnicas Lapidarias Prehispánicas*. México: INAH.

Mohar, Luz María. 1997. *Manos Artesanas del México Antiguo*. México: SEP, CONACYT.

Pastrana, Alejandro. 1998. *La exploración de la obsidiana en la Sierra de las Navajas*. Colección Científica, no. 383. México: INAH.

Pereira, Gregory. 2008. "La Materia de las Visiones. Consideraciones acerca de los espejos de pirita prehispánicos." *Diario de Campo*, (48): 123–136.

Pires-Ferreira, Jane W. 1975. "Formative Mesoamerican Exchange Networks with Special Reference to the Valley of Oaxaca." In *Prehistory and Human Ecology of the Valley of Oaxaca*, vol. 3, edited by Kent. V. Flannery. Museum of Anthropology Memoirs No. 7. Ann Arbor: University of Michigan.

Pires-Ferreira, Jane W., and Billy Joe Evans. 1978. "Mössbauer Spectral Analysis of Olmec Iron Ore Mirrors: New Evidence of Formative Period Exchange Networks." In *Cultural Continuity in Mesoamerica*, ed. David L. Browman, 101–154. The Hague, Netherlands: Mouton. http://dx.doi.org/10.1515/9783110807776.101.

Sahagún, Fray Bernardino de. 1989. *Historia General de la Cosas de la Nueva España.* México: Alianza Editorial Mexicana, Fondo de Cultura Económica.

Schele, Linda, and Mary E. Miller. 1986. *The Blood of Kings: Dynasty and Ritual in Maya Art.* New York: George Braziller.

Schiffer, Michael B. 1992. *Technological Perspectives on Behavioral Change.* Tucson: University of Arizona Press.

Schiffer, Michael B., James M. Skibo, Tamara C. Boelke, Mark A. Neupert, and Meredith Aronson. 1994. "New Perspectives on Experimental Archaeology: Surface Treatments and Thermal Response of the Clay Cooking Pot." *American Antiquity* 59 (2): 197–217.

Sepúlveda y Herrera, Maria Teresa. 2003. "La Matrícula de Tributos." *Arqueología Mexicana* (14): 1–85.

Smith, A. Ledyard, and Alfred V. Kidder. 1951. *Excavations at Nebaj, Guatemala.* Publication 594. Washington, DC: Carnegie Institution of Washington.

Taube, Karl A. 1992. "The Iconography of Mirrors at Teotihuacan." In *Art, Ideology, and the City of Teotihuacan*, ed. Janet Berlo, 169–204. Washington, DC: Dumbarton Oaks Research Library and Collection.

Thompson, J. Eric S. 1939. *Excavations at San Jose, British Honduras.* Publication no. 506. Washington, DC: Carnegie Institution of Washington.

Tringham, Ruth. 1978. "Experimentation, Ethnoarchaeology, and the Leapfrog in Archaeological Methodology." In *Explorations in Ethnoarchaeology*, ed. Richard A. Gould, 169–199. Albuquerque: University of New Mexico Press.

Turner, Margaret H. 1992. "Style in Lapidary Technology: Identifying the Teotihuacan Lapidary Industry." In *Art, Ideology, and the City of Teotihuacan*, ed. Janet Berlo, 89–112. Washington, DC: Dumbarton Oaks Research Library and Collection.

Valle, Perla. 1993. *Memorial de los indios de Tepetlaóztoc o Códice Kinghborough.* México: INAH.

Velázquez, Adrián. 2007. *La producción especializada de los objetos de concha del Templo Mayor de Tenochtitlan.* México: INAH.

Velázquez, Roberto. 2008. "Experimentación sobre cortes finos de rocas preciosas: una tecnología lapidaria milenaria poco conocida y estudiada." Accessed June 15, 2009. http://www.geocities.com/iztaccihuatlo8/lapidaria/cortes.html.

Weigand, Phil. 1995. "Mineria prehispánica en las regiones noroccidentales de Mesoamérica con énfasis en la turquesa." In *Arqueología del Occidente y Norte de México*, edited by Eduardo Williams and Phil Weighand, 115–138. Zamora: El Colegio de Michoacán.

Weigand, Phil. 1997. "La Turquesa." *Arqueología Mexicana* 27:26–36.

Widmer, Randolph J. 1991. "Lapidary Craft Specialization at Teotihuacan: Implication for Community Structure at 33:S3W1 and Economic Organization in the City." *Ancient Mesoamerica* 2 (1): 131–147. http://dx.doi.org/10.1017/S0956536100000444.

Woodward, Arthur. 1941. "Hohokam Mosaic Mirrors." *Los Angeles County Museum Quarterly* 1 (4): 6–11.

Zamora, Fabian Marcelo. 2002a. "La industria de la pirita en el sitio de Aguateca durante el periodo Clásico Tardío." Licenciatura Thesis, Department of Archaeology, Universidad del Valle de Guatemala, Guatemala.

Zamora, Fabian Marcelo. 2002b. "La industria de la pirita en el sitio Clásico Tardío de Aguateca." In *XV Simposio de Investigaciones Arqueológicas en Guatemala 2001*, ed. J. P. Laporte, H. Escobedo, and B. Arroyo, 695–708. Guatemala: Museo Nacional de Arqueología y Etnología.

3

Among the great universe of artifacts made in the pre-Hispanic world, mirrors or reflecting surfaces made of hematite, obsidian, or pyrite have begun to receive much more attention from researchers (Blainey 2007; Gallaga 2001, 2009; Pereira 2008; Salinas 1995). Due to their aesthetic beauty, but more for the complexity and skill required in the manufacture of mirrors, researchers agree that the hours invested in the production of one pyrite-encrusted mirror could have been enormous and, most likely, was one of the most complex items produced by pre-Hispanic artisans (Blainey 2007; Di Peso 1974; Furst 1966; Gallaga 2001, 2009; Kidder et al. 1946; Pereira 2008; Smith and Kidder 1951; Woodward 1941; Zamora 2002a, 2002b). Accordingly, it is appropriate that Kirchhoff (1943: 92–107) considers them as one of the many attributes or elements that define the Mesoamerican cultural region.

Even though researchers have previously recognized the difficulty of manufacturing mirrors and the importance of such items as prestige and magic-religious objects, there are very few studies about their manufacturing techniques and production organization. Because of this gap in the literature, we performed preliminary research on the technology of pyrite inlays from four Mesoamerican sites: Chiapa de Corzo and Tenam Puente (Chiapas), Teotihuacan (State of Mexico), and Tenochtitlan (Mexico City). The objective of this study was to identify potential tools and techniques employed in the manufacture of pyrite mirrors and inlays. This

Manufacturing Techniques of Pyrite Inlays in Mesoamerica

Emiliano Melgar,
Emiliano Gallaga M.,
and Reyna Solis

DOI: 10.5876/9781607324089.c003

technological analysis is based on experimental archaeology, as well as on the characterization of manufacturing traces with optic microscopy (OM) and scanning electron microscopy (SEM). The results obtained allow us to identify with great accuracy the pre-Hispanic tools employed in the manufacture of such objects and to distinguish different technological traditions that may reflect distinct styles indicating an item's regional or cultural origins.

PREVIOUS RESEARCH ON MANUFACTURING TECHNIQUES FOR PYRITE-MOSAIC MIRRORS

A lot of the previous research about pre-Hispanic mirrors had been focused on the symbolic meaning of such items, such as their morphology, trade, and use, but there have been very few works that study the manufacturing techniques or production of mirrors (Blainey 2007; Fialko 2000; Gallaga, 2001, 2009; Kidder et al. 1946; Mata Amado 2003; Nelson et al. 2009; Pereira 2008; Pires-Ferreira 1975; Pires-Ferreira and Evans 1978; Sugiyama 1992; Taube 1992; Zamora, 2002a, 2002b). When one of these items is found in an archaeological context, generally the researchers assume it to be an imported object and propose that its manufacture required high amounts of invested work time. These ideas are a consequence of the lack of knowledge on how they were made (Smith and Kidder 1951: 44). The few works that do address this issue propose the use of emery (aluminum oxide), hematite, sand, and ocher to abrade and polish the pyrite mosaics, based on the scarce written references about the tools employed by the pre-Hispanic lapidary artisans (Lunazzi 1996: 3). Even more, some researchers propose the use of a compass to ensure a perfect circular stone base for the mirrors (Smith and Kidder 1951: 44). However, these are just suppositions, because to confirm such hypotheses it is necessary to perform a technological analysis of the manufacturing traces and a comparison with the proposed tools. This procedure, as documented herein, is a classic implementation of experimental archaeology.

Even more rare are analyses of the pyrite plaques' production, but some exceptions do exist. Among the studies that stand out are those concerning the sites of Aguateca and Cancuén. At the first site, pyrite items were found in an elite residential and administrative context where nine nodules of raw material, 77 pieces in process or recycled, and 484 finished objects were identified (455 mosaic plaques, 19 tridimensional shapes, and 10 plaques) (Zamora 2002b: 696). Although the analysis is based on descriptions of the pieces, unfortunately it only mentions that the items were rounded on the corners, and cut and polished on one or two sides (Zamora 2002b: 700), without any proposition regarding how they were made. At Cancuén, pyrite items were recovered at domestic units of non-elite groups. The excavated material consisted of raw pieces, pieces in process, and finished plaques ready to be assembled onto mirror bases. Researchers also mention that there are stone bases associated with chert and quartzite tools for smoothing and polishing the pyrite items (Kovacevich et al. 2002: 337; 2004: 885). However, very few comments are focused on the tools, because most of the attention is paid to the organization of the production of such items. In general, the researchers propose that the first steps in the production of the mirrors were carried out by non-elite domestic groups, while in order to control and monopolize both the production and consumption of these goods, the last steps were completed by specialized artisans of the elite classes (Kovacevich et al. 2002: 343; 2004: 888; Kovacevich 2007: 86–90).

TECHNOLOGICAL ANALYSIS: A PROPOSAL FOR PYRITE ITEMS

Due to the lack of information on lapidary technology from historical sources and the fact that very little research on the manufacture of pyrite objects has extended beyond the hypothetical level, further clarifications require the application of experimental archaeology methods. In this regard, some preliminary steps have thus far been completed to reproduce a single pyrite plaque. While these pursuits began as a hobby, they produced some interesting results (Ekholm 1973; Gallaga 2009; Zamora 2002a).

In accord with the experimental archaeology paradigm, it is assumed that in human societies most of the activities are regulated. This means that all of the artifacts produced or used follow a particular pattern that gives them specific characteristics (Ascher 1961: 807; Gándara Vázquez 1990: 51). Along this line of thinking, the employment of similar tools and techniques, following certain patterns, will produce identical pieces in comparison with the archaeological ones (Ascher 1961: 793; Coles 1979: 171; Lewenstein 1987: 7; Velázquez Castro 2007: 52–54). Based on this theory and following the uniformity criteria for the use of a particular tool made of a certain material, used in a standard procedure, and under particular conditions, we can further assume that this will produce specific and characteristic traces that allow us to identify and distinguish each of them (Binford 1977: 7; Tringham 1978: 180; Velázquez Castro 2007: 23).

With this model in mind and with the objective to fill holes in the lack of knowledge about the pre-Hispanic methods and technology used in lapidary object production, the Workshop of Experimental Archaeology in Lapidary Project (figure 3.1) was created in 2004 at the Templo Mayor Museum inside the

Lapidary of Templo Mayor: Styles and Technological Traditions (Melgar Tísoc 2009a; Melgar Tísoc et al. 2010; Melgar Tísoc et al. 2011; Melgar Tísoc and Solís Ciriaco 2009). Since its creation, the workshop has been able to reproduce the different types of modifications that the archaeological pieces exhibit (figure 3.1 and table 3.1). To get to this level of knowledge, we reproduced the tools and procedures described in the historic sources (Durán 1967; Sahagún 1956), the tools found in archaeological contexts (Feinman and Nicholas 1993; Hohmann 2002; Lewenstein 1987; Moholy-Nagy 1997; Pastrana 1998), and those proposed by researchers (Athie Islas 2001; Charlton 1993; Digby 1964; Gazzola 2007; Gómez Chávez 2000; Langenscheidt 2006, 2007; Mirambell 1968; Semenov 1957).

The obtained traces are systematically compared under controlled parameters with those found on the archaeological items by: simple eye; magnifying glass of 20x; OM at 10x, 30x, and 63x; and SEM at 100x, 300x, 600x, and 1,000x. In the case of the SEM analysis, we employed HV mode, 10 mm of work distance, SEI signal, voltage of 20kV, and spotsize of 47. Following the methodology proposed by Velázquez Castro (2007) for the study and description of shell items adapted for lapidary objects, the manufacturing traces are described, focusing on the direction and size of bands and lines, and on the object's texture and rugosity. The optical microscope that was employed is located at the Workshop of Experimental Archaeology in Shell at the Templo Mayor Museum, and the SEM we employed is found at the Electron Microscopy Laboratory of the Subdivision of Laboratories and Academic Backup of INAH.

ANALYZED MATERIALS

In this study we analyzed archaeological pyrite inlays from four Mesoamerican sites and samples from two pyrite mirrors from the state of Chiapas, Mexico. One of them has rectangular inlays and shape (figure 3.2a). It was found inside a tomb in mound 11 of Chiapa de Corzo and it was dated to around the Middle Preclassic period (700–500 BC) (Bachand and Lowe 2011). The second mirror has a circular shape (figure 3.2b). It came from tomb 10 of building 21 at Tenam Puente and it was dated to the Late Classic period (AD 600–900) (Martínez del Campo Lanz 2010: 76–77). From each of them, four pyrite plaques were selected. In order to perform the analysis of the manufacturing traces, these plaques were chosen based on their well-preserved surface and sides, and for the minimal amounts of oxide present on their surfaces.

Also, we analyzed 39 inlays: three circular pieces from Teopancazco (figure 3.2c), a neighborhood center located at Teotihuacan and dated to the Classic period (AD 200–600) (Manzanilla 2009), and 36 semispherical pyrite

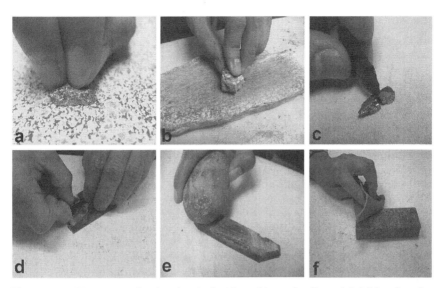

FIGURE 3.1. *Experimental archaeology in lapidary objects: abrading with (a) basalt and (b) sandstone, cutting with (c) obsidian and (d) chert flakes, and (e) polishing with chert pebbles and (f) brightening with leather (photos by Emiliano Melgar).*

TABLE 3.1. Types of modifications and tools employed

Modification	Tools
Abrading	Basalt, andesite, rhyolite, sandstone, limestone, granite, and slate, adding water and occasionally sand.
Cutting	Sand, water, and strips of hide. Chert and obsidian lithic tools.
Drilling	Abrasives (sand, volcanic ash, obsidian dust, and chert dust), worked with reed sticks and adding water. Chert and obsidian lithic tools.
Incising	Chert and obsidian lithic tools.
Polishing and brightening	Polishing with abrasives, water, and leather pieces. Brightening with pieces of leather. Combining both techniques.

incrustations from the Templo Mayor of Tenochtitlan in the Postclassic period (AD 1325–1521); seven from the chert scepter representing the *Xiuhcóatl*, a snake decorated with a turquoise mosaic and pyrite from chamber III (figure 3.2d) and dated to the construction stage IVa (AD 1440–1469); 26 samples were the eyes from 13 cranium masks (figure 3.2e) from offerings 6, 11, 13, 17, and 20; and the last three samples were from the eyes of a travertine mask

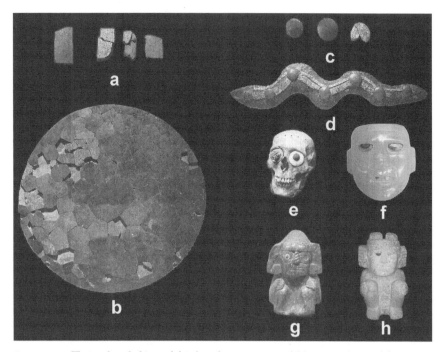

FIGURE 3.2. *The analyzed objects: (a) inlays from a mirror of Chiapa de Corzo, (b) inlays from a mirror of Tenam Puente, (c) inlays from Teopancazco at Teotihuacan, (d) Xiuhcoatl scepter of chert with turquoise and pyrite inlays, (e) travertine mask with shell and pyrite inlays, (f) skull mask with shell and pyrite inlays, (g) Tlaloc sculpture with pyrite inlays, and (h) Xiuhtecuhtli sculpture with pyrite inlays, all of them from the Great Temple of Tenochtitlan (photos by Emiliano Melgar).*

of offering 82 (figure 3.2f), the sculpture of *Tláloc* made of serpentine (figure 3.2g), and the sculpture of *Xiuhtecuhtli* made of travertine (Figure 3.2h) from Chamber II, all dated to the construction stage IVb (AD 1469–1481).

RESULTS OBTAINED

In general terms, all the pyrite incrustations from mirrors showed smooth and polished surfaces crossed by some straight diffuse lines at both 10x (figure 3.3) and 100x (figure 3.4). In the particular case of the Chiapa de Corzo samples (figure 3.3a), there are diffuse straight bands of 33 μm (figure 3.4.a), similar to the experimental traces produced by the abrading with rhyolite *metates* and polished with leather (figures 3.3b and 3.4b). In contrast, the samples from

Tenam Puente (figure 3.3c) showed diffuse flattened bands of 100 μm with a rugged texture (figure 3.4c). This pattern is similar to those experimental traces obtained by the abrading with basalt metates and sand as abrasive, and the brightening with leather (figures 3.3d and 3.4d). It is important to mention that in both mirrors the observed bands were more diffuse on the frontal surface than on the back surface, probably because the first one had brightening and the second was in contact with the base and the adhesive.

In the case of the Teopancazco circular samples, all showed the flattened and polished surface (figure 3.3e) crossed by diffuse bands of 66 μm (figure 3.4e) and fine parallel-straight bands between 2 and 5μm. This pattern is similar to those produced by the experimental abrading with andesite *metates*, polishing with chert nodules, and brightening with leather (figures 3.3f and 3.4f).

Finally, all the pyrite samples from the Templo Mayor of Tenochtitlan showed flattened and polished surfaces crossed by several parallel-straight bands (figure 3.3g) that measure 100 μm (figure 3.4g). This pattern is similar to those obtained by the experimental abrading with basalt *metates* and polishing with leather (figures 3.3h and 3.4h).

The analyses of the edges provide information about how the pieces could be made. The samples present the following traces: the mirrors have straight edges with a rugged texture (figures 3.5a and 3.5b), crossed by fine lines between 0.5 and 2 μm width (figures 3.6a and 3.6b), similar to those obtained by experimental cutting with obsidian flakes or blades (figures 3.5d and 3.6d). In contrast, the edges of the circular incrustations are flattened and polished (figure 3.5c), crossed by parallel-straight bands of 2–5 μm (figures 3.6c and 3.6d). This pattern is similar to the experimental cutting with chert tools (figures 3.5e and 3.6e). Finally, it is interesting that all of the semispherical samples do not show edges made by cutting tools because the shape was made only by abrading.

DISCUSSION

As we mention at the beginning of this chapter, the majority of previous research into pyrite objects, especially mirrors, has concentrated on the symbolic use of this material, its morphological variability, its trade, and its use/consumption context (Blainey 2007; Fialko 2000; Gallaga 2001; Kidder et al. 1946; Mata Amado 2003; Pereira 2008; Pires-Ferreira 1975; Pires- Ferreira and Evans 1978; Sugiyama 1992; Taube 1992; Zamora, 2002a, 2002b). These publications contrast with the relative lack of studies on manufacturing techniques. Unfortunately, the majority of these scholarly treatments of mirrors recapitulate the descriptions made in historical sources and/or guess at the

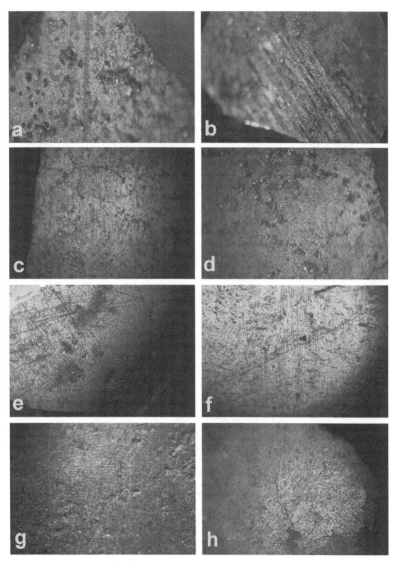

FIGURE 3.3. *Analyses of surfaces (all at 10x): Chiapa de Corzo (a) inlay and (b) experimental abrading with rhyolite and brightening with leather. Tenam Puente (c) inlay and (d) experimental abrading with basalt and sand and brightening with leather. Teopancazco (e) inlay and (f) experimental abrading with andesite, polishing with chert pebbles, and brightening with leather. Tenochtitlan (g) inlay and (h) experimental abrading with basalt and brightening with leather (photos by Emiliano Melgar).*

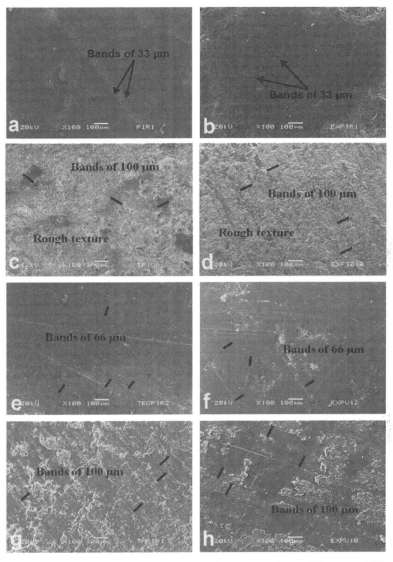

FIGURE 3.4. *Analyses of surfaces (all at 100x): Chiapa de Corzo (a) inlay and (b) experimental abrading with rhyolite and brightening with leather. Tenam Puente (c) inlay and (d) experimental abrading with basalt and sand and brightening with leather. Teopancazco (e) inlay and (f) experimental abrading with andesite, polishing with chert pebbles, and brightening with leather. Tenochtitlan (g) inlay and (h) experimental abrading with basalt and brightening with leather (photos by Emiliano Melgar and Gerardo Villa).*

FIGURE 3.5. *Analyses of edges (10x): (a) Chiapa de Corzo inlay, (b) Tenam Puente inlay, (c) Teopancazco inlay, and experimental cutting with (d) obsidian and (e) chert flakes (photos by Emiliano Melgar).*

possible tools employed without any experiment to confirm these suppositions (Blainey 2007: 177–179; Ekholm 1973: 134; Gallaga 2009; Kovacevich et al., 2002: 343, 2004: 885–888; Lunazzi 1996: 3; Zamora 2002b: 696–700). To fill this gap, the experimental archaeology analyses performed by the lapidary workshop at Templo Mayor allowed us to identify the tools employed on each modification of pyrite samples with high accuracy. With these procedures, it is possible to confirm or refute the set of tools proposed by researchers in previous analyses and to "reconstruct" the production sequence of any single

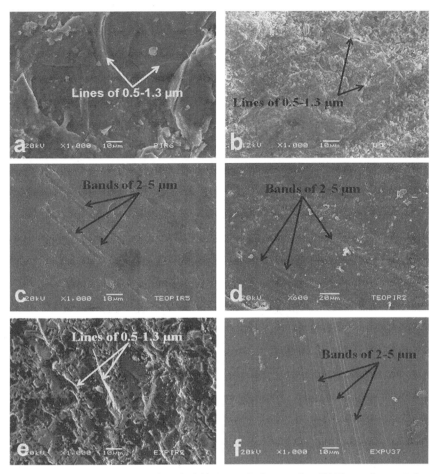

FIGURE 3.6. *Analyses of edges (100x): (a) Chiapa de Corzo inlay, (b) Tenam Puente inlay, (c) Teopancazco inlay, and experimental cutting with (d) obsidian and (e) chert flakes (photos by Emiliano Melgar).*

pyrite object. In this way, it is possible to recover a lot of information about the manufacture of these objects and to demonstrate that the direct evidences of production, such as raw material, pieces in process, refuse, and associated tools, found in primary (workshops) or secondary (garbage pit and construction fill) contexts, are not the only ways to know the manufacture process in detail. As we showed briefly in this study, it is possible to recover almost all the steps of the manufacture process from a single finished piece regardless of its archaeological context.

Another result of our analysis gives us a better idea about the organization of the production of these types of objects. Previous research with this purpose is based on the four specialization artisans' parameters proposed by Costin (1991): context (dependent or independent from the elite control), concentration (centralization or dispersion of the work groups), scale (size of the work groups), and intensity (half- or full-time invested in work). With these parameters in mind, two scenarios were presented: on one side the diverse evidence for pyrite production, especially in domestic units of non-elite groups, and the findings of finished pieces in elite contexts, indicates half-time specialization activities performed by artisans who were non-elite members but who produced pyrite goods for elites (Kovacevich 2007: 86–90; Kovacevich et al. 2002: 343). On the other hand, when the production evidence and finished objects are found only in elite and/or administrative residential contexts, researchers have proposed that the production was carried out by elite artisans and members of such social classes, because the final manufacturing stages required esoteric knowledge and special rituals (Inomata 2001: 321; 2007: 134–135; Kovacevich, et al. 2002: 344; 2004: 887–888; Kovacevich 2007: 74–76, 90). In the proposed scenarios for the manufacture of these objects, the most important indicators that the researchers employed focused on evidences of production, the workshops, the deposition of the materials, and the differences in the spatial distribution between evidence of its production and the pyrite mirrors. However, almost nothing is mentioned about the size of the working force used for their production. In addition, it is assumed that the artisans are half-time specialists and have low-volume production because the evidence was found in domestic contexts and mixed with other items not related to pyrite-production objects, such as domestic tools (Kovacevich et al. 2004: 887–888).

In the particular case of the production process, the results from this analysis show evidence that the objects were made with a strong morphological standardization of production, especially in the tool technology. If we add to this information the contexts of the pyrite samples (tombs and offerings), it allows us to think that the workforce engaged in the manufacture of these items was centralized and under the control of elites. In addition, the homogeneity of shapes and tools used for the manufacture of pyrite items at the four different sites offers strong evidence to indicate that the workgroups were small, concentrated, and/or centralized. In these circumstances, the control and supervision of the manufacturing process by master artisans or elite members is facilitated (Costin 1991: 16, 40; Velázquez Castro 2007: 19–20). Of course these results are preliminary and more research in the future is necessary.

In evaluating the relative intensification or the hours invested to produce a single pyrite mirror, several researchers have already presumed this investment to be high, in that it must have required a considerable amount of time (Blainey 2007; Di Peso 1974; Furst 1966; Gallaga 2001; Kidder et al. 1946; Pereira 2008; Smith and Kidder 1951; Woodward 1941; Zamora 2002a, 2002b). Thus, it was necessary to perform an experiment to attempt to reproduce the manufacturing steps. Emiliano Gallaga (2009; see chapter 2, this volume) performed a first attempt using a piece of sandstone to reproduce the base of a mirror and working a piece of pyrite to make a plaque: The manufacture of the stone base of 9 cm in diameter by 8 mm thick took 90 hours to complete: three hours to work a preform with stone hammers, 51 hours to abrade the piece with sand and water on a basalt stone metate, eight hours to bevel the sides of it, and 28 hours to make two conical holes at the center of it with quartz stones mounted on reed sticks. The manufacture of a single pyrite plaque of 1 cm^2 by 4 mm thick took 41 hours, employing stone tools to reduce pyrite nodules into small pieces and then by abrading the pieces with sand and water on a basalt metate. If we consider that a mirror is composed of 20 to 30 pyrite plaques of similar dimensions, the hours involved to produce this many plaques would increase considerably.

Although these are preliminary results of temporal estimations for the manufacture of a pyrite mirror 9 cm in diameter with a mosaic of 20 to 30 pyrite plaques, to produce the entire mirror would have required an estimated 900–1300 hours, or 110–160 days of work, for a single person. This amount of time could be reduced if people were very skilled after lots of practice or if multiple people were working on the same object and divided the work for different steps of manufacture. Also, if we take into account the time estimated to produce the mirror, the acquisition of the raw material, and the skills required by the artisans, we could infer that the artisans must have been full-time specialists. On the other hand, the analysis of the pyrite samples illustrates that the tools and steps of production were not always the most efficient, especially those used for abrading the pieces. In this line of evidence, only the samples from the site of Tenam Puente showed evidence of using sand over basalt. In contrast, the samples from Chiapa de Corzo, Teotihuacan, and Templo Mayor only used basalt, andesite, and/or rhyolite metates without any type of abrasives.

Such results indicate that the artisans who worked these items did not attempt to manufacture more items in the least time possible, probably due to the fact that the value of the pieces includes the amount of time invested in them (Melgar Tísoc 2009b: 261–263; Widmer 2009: 182); because mirrors are prestige items, they do not require a sparing of time or raw material (Shimada 1994: 25, Velázquez Castro 2007: 18). In addition, the selection of certain tools is not determined most of the time by the geographical location or access to material; sometimes it depends on cultural, traditional, ideological, religious, or political patterns (Gosselain 1992: 580; Lemonnier 1986: 153; Melgar Tísoc 2009b: 5; Pfaffenberger 1988: 249; Schiffer 1992: 51; Velázquez Castro 2007: 22).

In this sense, it is interesting to mention the variability of tools used at each of the sites analyzed. Perhaps this technological difference reflects distinct manufacture traditions or regional technological styles unique to the Zoque, Teotihuacan, Maya, and Aztec people. This idea is based on the statement that every culture or social group has its particular and characteristic ways to create their objects, whereby every style can be considered as the systematic selection and established repetition of how to manufacture an object, like a pyrite mirror, in a specific space and time frame (Carr 1995: 166; Velázquez Castro 2007: 21–22). In this way, the introduction of technological variability into the established style analysis allows us to go deeper into the ways that a culture expresses itself through the manufacture of an object and how this marks them with a particular identity or method of production (Gosselain 1992: 583; Wobst 1977: 321).

With these ideas in mind, future phases of analysis will be required to compare the manufacturing traces and tools used in other sites and on lapidary objects from the region under study to attempt to identify and distinguish technological traditions across space and time. For example, the tools identified in the pieces from Teotihuacan (figures 3.7a–d) and Tenochtitlan (figures 3.7e–h) were also found in the production of other lapidary objects from those sites (Melgar Tísoc 2009a; Melgar Tísoc and Filloy Nadal 2008; Melgar Tísoc and Solís Ciriaco 2009; Melgar Tísoc, Solís Ciriaco, and Ruvalcaba Sil 2011). Some of them identify as "Teotihuacan style" pieces (such as trapezoidal pendants and anthropomorphic figurines) recovered from different temples, pyramids, or neighborhoods at Teotihuacan (Turner 1992) or "Mexica manufactures" (such as Tezcatlipoca's circular pectorals called *anáhuatl*, Techalotl's cylindrical scepters, Xipe Totec's rectangular nose plugs, figurines representing Tláloc, and mosaic representations of Huitzilopochtli, Mixcóatl, and Tlahuizcalpantecuhtli) found in the offerings from the Great Temple of Tenochtitlan (Melgar Tísoc 2009a; Melgar Tísoc, Solís Ciriaco, and Ruvalcaba Sil 2011; Velázquez Castro 2007). These data support the argument that the technological skills for lapidary work in both sites are characteristic of the Teotihuacan and Mexican artisans respectively. In the case of the samples from Chiapa de Corzo and Tenam Puente,

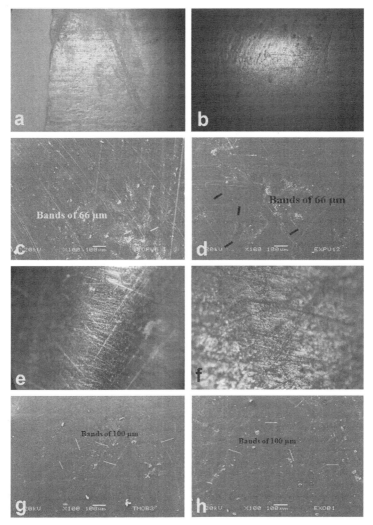

FIGURE 3.7. *Technological comparison with pieces from Teotihuacan and Tenochtitlan: (a) surface of a travertine vase from Teotihuacan, and (b) experimental abrading with andesite, polishing with chert pebbles, and brightening with leather; both at 10x. (c) Surface of a greenstone figurine from Teotihuacan, and (d) experimental abrading with andesite, polishing with chert pebbles, and brightening with leather; both at 100x. (e, g) Surface of an obsidian scepter from Tenochtitlan, 10X and 100x, respectively, and (f, h) experimental abrading with basalt and brightening with leather, 10x and 100x, respectively (photos by Emiliano Melgar and Gerardo Villa).*

it is necessary to conduct more research on other objects to establish local or regional lapidary manufacturing traditions.

CONCLUDING REMARKS

The technological research on lapidary objects, carried out through experimental archaeology and the characterization of the manufacturing traces by means of optical microscope and scanning electron microscope allows us to know with high precision the tools employed in each modification that the item presented.

In this research, 43 samples from four Mesoamerican archaeological sites were analyzed (Chiapa de Corzo, Tenam Puente, Teotihuacan, and Tenochtitlan). The preliminary results permit us to conclude that each site used different tools to manufacture or work pyrite items, and that each site presented its own morphological and technological standardization of production. This scenario allows us to infer some aspects of the organization of production, such as that the size of the workgroup must have been small, concentrated, and/or centralized, and directly controlled and/or supervised by master artisans or elite members.

By the performance of a preliminary experimental archaeology project, with the result of a stone base and a single pyrite plaque that took more than 100 effective hours of work, we estimate an average of 900–1,300 hours, or 110–160 days, of work for a single artisan to make a single pyrite mirror (Gallaga 2009). This amount of time could be reduced if people were very skilled after lots of practice or if multiple people were working on the same object and divided the work of the steps of manufacture. In addition, the manufacturing patterns identified are different between the four sample sites. In the case of the Teotihuacan and Tenochtitlan pieces, we identified similar technological patterns in other lapidary objects, making it possible to establish their local manufacturing tradition, and perhaps, the workshop signature.

Finally, the technological analysis of the tools used by the characterization of the manufacturing traces indicates that the pre-Hispanic artisans did not employ the most efficient tools or procedures to produce a mirror. Probably this is due to the fact that they were more interested in increasing the value of the items based on the labor invested and not in mass production. Also, it is important to keep in mind that the pyrite mirrors are not simple and common objects; indeed, these were expensive and prestigious items intended as the property of elites.

ACKNOWLEDGMENTS

This work could not be possible without the help and collaboration of Adrián Velázquez from the Research Department of Templo Mayor Museum; Fernando Carrizosa, María Elena Cruz, and Marcela Castaño from the Storehouse of Cultural Resources of the Templo Mayor Museum; Gerardo Villa from the Electron Microscopy Laboratory of INAH; and the members of the Workshop of Experimental Archaeology in Lapidary Objects (Mauricio Valencia, Isaac Ramírez, Mijaely Castañón, Hervé Monterrosa, and Edgar Pineda). Much thanks also to Bruce Bachand and Lynneth Lowe, and Emiliano Gallaga from the Chiapa de Corzo Archaeological Project; to Lalo Jacinto, director of the Tenam Puente Archaeological Project; to Linda R. Manzanilla, director of the project Teotihuacan: Elite and Government, for facilitating the pyrite samples for the analysis; and to Laura Filloy from the Restoration Laboratory at the National Museum of Anthropology for the comparative lapidary samples from the Pyramid of the Moon Project.

REFERENCES

Ascher, Robert. 1961. "Experimental Archaeology." *American Anthropologist* 63 (4): 793–816. http://dx.doi.org/10.1525/aa.1961.63.4.02a00070.

Athie Islas, Ivonne. 2001. "La Obsidiana del Templo Mayor de Tenochtitlan." Bachelor's thesis, Escuela Nacional de Antropología e Historia, Mexico City.

Bachand, Bruce R., and Lynneth S. Lowe. 2011. "Chiapa de Corzo y los olmecas." *Arqueología Mexicana* 107: 74–83.

Blainey, Marc G. 2007. "Surfaces and Beyond: The Political, Economic, and Ideological Significance of Ancient Maya Iron Ore Mirrors." Master's Thesis, Trent University, Peterborough, Ontario, Canada.

Binford, Lewis R. 1977. "General Introduction." In *For Theory Building in Archaeology: Essays on Faunal Remains, Aquatic Resources, Spatial Analysis, and Systemic Modeling*, by Lewis R. Binford, 1–10. Albuquerque: Academic Press.

Carr, Christopher. 1995. "Building a Unified Middle-Range Theory of Artifact Design." In *Style, Society and Person*, ed. Christopher Carr and Jill E. Neitzel, 151–170. New York: Plenum Press. http://dx.doi.org/10.1007/978-1-4899-1097-4_6.

Charlton, Cynthia Otis. 1993. "Obsidian as Jewelry: Lapidary production in Aztec Otumba, México." *Ancient Mesoamerica* 4 (2): 231–243. http://dx.doi.org/10.1017/S0956536100000924.

Coles, John. 1979. *Experimental Archaeology*. London: Academic Press.

Costin, Cathy Lynne. 1991. "Craft Specialization: Issues in Defining, Documenting, and Explaining the Organization of Production." In *Archaeological Method and Theory*, vol. 3. ed. Michael B. Schiffer, 1–56. Tucson: University of Arizona Press.

Digby, Adrian. 1964. *Maya Jades*. London: The Trustees of the British Museum.

Di Peso, Charles. 1974. *Casas Grandes: A Fallen Trading Center of the Grand Chichimeca*. vol. 1. Flagstaff, AZ: Amerind Foundation.

Durán, Fray Diego. 1967. *Historia de las indias de Nueva España e islas de tierra firme*. México: Porrúa.

Ekholm, Gordon F. 1973. "The Archaeological Significance of Mirrors in the New World." In *Atti del XL Congresso Internazionale degli Americanisti, Roma-Genova, 1972* vol. 1: 133–135. Genova: Casa Editrice Tilgher.

Feinman, Gary M., and Linda M. Nicholas. 1993. "Shell-Ornament Production in Ejutla: Implications for Highland-Coastal Interaction in Ancient Oaxaca." *Ancient Mesoamerica* 4 (1): 103–119. http://dx.doi.org/10.1017/S095653610000081X.

Fialko, Vilma. 2000. "El espejo del Entierro 49; morfología y texto jeroglífico." In *El sitio maya de Topoxté: Investigaciones en una isla del lago Yaxhá, Petén, Guatemala*, ed. Wolfgang W. Wurster, 144–149. Mainz am Rhein, Germany: Verlag.

Furst, Peter T. 1966. "Shaft Tombs, Shell Trumpets and Shamanism: A Culture-Historical Approach to Problems in West Mexican Archaeology." PhD diss., University of California, Los Angeles.

Gallaga M., Emiliano. 2001. "Descripción y análisis de los espejos de pirita del sitio de Snaketown, AZ, EU y su relación con Mesoamérica." Paper presented at the XXVI Mesa Redonda de Antropología, Zacatecas, México, July 29–August 3.

Gallaga M., Emiliano. 2009. "La manufactura de los espejos de pirita: Una experimentación." Paper presented at the 53 International Congress of Americanists, México City, July 19–24.

Gándara Vázquez, Manuel. 1990. "La analogía etnográfica como heurística: Lógica muestreal, dominio etnográfico e historicidad." In *Etnoarqueología: Primer Coloquio Bosch-Gimpera*, edited by Yoko Sugiura and Mari Carmen Serra Puche, 43–82. México: UNAM.

Gazzola, Julie. 2007. "La producción de cuentas en piedras verdes en los talleres lapidarios de La Ventilla, Teotihuacan." *Arqueología* 36: 52–70.

Gómez Chávez, Sergio. 2000. "La Ventilla: un barrio de la antigua ciudad de Teotihuacan: Exploraciones y resultados." Bachelor's thesis, Escuela Nacional de Antropología e Historia, Mexico City.

Gosselain, Olivier P. 1992. "Technology and Style: Potters and Pottery among Bafia of Cameroon." *Man* 27 (3): 559–583. http://dx.doi.org/10.2307/2803929.

Hohmann, Bobbi M. 2002. "Preclassic Maya Shell Ornament Production in the Belize Valley, Belize." PhD diss., University of New Mexico, Albuquerque.

Inomata, Takeshi. 2001. "The Power and Ideology of Artistic Creation: Elite Craft Specialists in Classic Maya Society." *Current Anthropology* 42 (3): 321–349. http://dx .doi.org/10.1086/320475.

Inomata, Takeshi. 2007. "Classic Maya Elite Competition, Collaboration, and Performance in Multicraft Production." In *Craft Production in Complex Societies: Multicraft and Producer Perspectives*, ed. Izumi Shimada, 120–133. Salt Lake City: University of Utah Press.

Kidder, Alfred V., Jesse D. Jennings, and Edwin M. Shook. 1946. *Excavation at Kaminaljuyu, Guatemala*. Washington, DC: Carnegie Institution.

Kirchhoff, Paul. 1943. "Mesoamérica: Sus límites geográficos, composición étnica y caracteres culturales." *Acta Americana* 1: 92–107.

Kovacevich, Brigitte. 2007. "Ritual, Crafting, and Agency at the Classic Maya Kingdom of Cancuén." In *Mesoamerican Ritual Economy: Archaeological and Ethnological Perspectives*, edited by E. Christian Wells and Karla L. Davis Salazar, 67–114. Boulder: University Press of Colorado.

Kovacevich, Brigitte, Tomás Barrientos, Michael Callaghan, and Karen Pereira. 2002. "La economía en el Reino Clásico de Cancuén: Evidencia de producción, especialización e intercambio." In *XV Simposio de Investigaciones Arqueológicas en Guatemala, 2001*, ed. J. P. Laporte, H. Escobedo, and B. Arroyo, 333–349. Guatemala: Museo Nacional de Arqueología y Etnología.

Kovacevich, Brigitte, Duncan Cook, and Timothy Beach. 2004. "Áreas de actividad doméstica en Cancuén: Perspectivas basadas en datos líticos y geo-químicos." In *XVII Simposio de Investigaciones Arqueológicas en Guatemala, 2003*, ed. J. P. Laporte, B. Arroyo, H. Escobedo, and H. Mejía, 876–891. Guatemala: Museo Nacional de Arqueología y Etnología.

Langenscheidt, Adolphus. 2006. "Los abrasivos en Mesoamérica." *Arqueología Mexicana* 80: 55–60.

Langenscheidt, Adolphus. 2007. "Lapidaria mesoamericana, una reflexión sobre los abrasivos posiblemente usados para trabajar los chalchihuites duros." *Arqueología* 36: 179–206.

Lemonnier, Pierre. 1986. "The Study of Material Culture Today: Toward an Anthropology of Technical Systems." *Journal of Anthropological Archaeology* 5 (2): 147–186. http://dx.doi.org/10.1016/0278-4165(86)90012-7.

Lewenstein, Suzanne M. 1987. *Stone Tools at Cerros: The Ethnoarchaeological and Use-Wear Evidence*. Austin: University of Texas Press.

Lunazzi, José J. 1996. "Olmec Mirrors: An Example of Archaeological American Mirrors." In *Trends in Optics*, ed. Anna Consortini, 411–421. London: Academic Press. http://dx.doi.org/10.1016/B978-012186030-1/50024-2.

Manzanilla, Linda. 2009. "Corporate Life in Apartment and *Barrio* Compounds at Teotihuacan, Central Mexico: Craft Specialization, Hierarchy and Ethnicity." In *Domestic Life in Prehispanic Capitals: A Study of Specialization, Hierarchy and Ethnicity*, ed. Linda R. Manzanilla and Claude Chapdelaine, 21–42. Ann Arbor: University of Michigan.

Martínez del Campo Lanz, Sofía. 2010. *Rostros de la divinidad: Los mosaicos mayas de piedra verde*. México: INAH.

Mata Amado, Guillermo. 2003. "Espejo de pirita y pizarra de Amatitlán." In *XVI Simposio de Investigaciones Arqueológicas en Guatemala, 2002*, edited by J. P. Laporte, B. Arroyo, H. Escobedo, and H. Mejía, 831–839. Guatemala: Museo Nacional de Arqueología y Etnología.

Melgar Tísoc, Emiliano Ricardo. 2009a. "Análisis tecnológico de los objetos de piedra verde del Templo Mayor de Tenochtitlan." Paper presented at the 53rd International Congress of Americanists, Mexico City, July 19–24.

Melgar Tísoc, Emiliano Ricardo. 2009b." La producción especializada de objetos de concha en Xochicalco." Master's thesis, Universidad Nacional Autónoma de México, Mexico City.

Melgar Tísoc, Emiliano Ricardo, and Laura Filloy Nadal. 2008. "Manufactura y restauración de una figurilla antropomorfica hecha de mosaico de piedra verde hallada en la Pirámide de la Luna, Teotihuacan, México." Paper presented at the XVII Internacional Materials Research Congress of the Academia Mexicana de Ciencia de Materiales, Cancún, August 17–21.

Melgar Tísoc, Emiliano Ricardo, and Reyna Beatriz Solís Ciriaco. 2009. "Caracterización de huellas de manufactura en objetos lapidarios de obsidiana del Templo Mayor de Tenochtitlan." *Arqueología* 42: 118–134.

Melgar Tísoc, Emiliano Ricardo, Reyna Beatriz Solís Ciriaco, and Ernesto González Licón. 2010. "Producción y prestigio en concha y lapidaria de Monte Albán." In *Producción de bienes de prestigio ornamentales y votivos de la América antigua*, ed. Emiliano Melgar Tísoc, Reyna Solís Ciriaco, and Ernesto González Licón, 6–21. Miami: Syllaba Press.

Melgar Tísoc, Emiliano Ricardo, Reyna Beatriz Solís Ciriaco, and José Luis Ruvalcaba Sil. 2011. "Del centro de barrio al complejo palaciego: los artesanos lapidarios y las tradiciones de manufactura locales y foráneas vistas desde Teopancazco y Xalla." Paper presented at the V Teotihuacan Round Table, Teotihuacan, October 23–28.

Mirambell, Lorena. 1968. *Técnicas Lapidarias Prehispánicas*. México: INAH.

Moholy-Nagy, Hattula. 1997. "Middens, Construction Fill, and Offerings: Evidence for the Organization of Classic Period Craft Production at Tikal, Guatemala." *Journal of Field Archaeology* 24 (3): 293–313. http://dx.doi.org/10.1179/00934699779 2208096.

Nelson, Zachary, Barry Scheetz, Guillermo Mata Amado, and Antonio Prado. 2009. "Composite Mirrors of the Ancient Maya: Ostentatious Production and Precolumbian Fraud." *The PARI Journal* 9: 1–16.

Pastrana, Alejandro. 1998. *La exploración de la obsidiana en la Sierra de las Navajas*. México: INAH.

Pereira, Gregory. 2008. "La Materia de las Visiones. Consideraciones acerca de los espejos de pirita prehispánicos." *Diario de Campo* 48:123–136.

Pfaffenberger, Bryan. 1988. "Fetishised Objects and Humanised Nature: Towards an Anthropology of Technology." *Man* 23 (2): 236–252. http://dx.doi.org/10.2307 /2802804.

Pires-Ferreira, Jane W. 1975. *Formative Mesoamerican Exchange Networks with Special Reference to the Valley of Oaxaca: Prehistory and Human Ecology of the Valley of Oaxaca*. Ann Arbor: University of Michigan.

Pires-Ferreira, Jane W., and Billy Joe Evans. 1978. "Mössbauer Spectral Analysis of Olmec Iron Ore Mirrors: New Evidence of Formative Period Exchange Networks." In *Cultural Continuity in Mesoamerica*, ed. David L. Browman, 101–154. The Hague: Mouton. http://dx.doi.org/10.1515/9783110807776.101.

Sahagún, Fray Bernardino de. 1956. *Historia General de las Cosas de la Nueva España*. México: Porrúa.

Salinas, Flores. 1995. *Tecnología y Diseño en el México Prehispánico*. México: UNAM.

Semenov, S. A. 1957. *Prehistoric Technology: An Experimental Study of the Oldest Tools and Artifacts from Traces of Manufacture and Wear*. London: Cory, Adams and MacKay.

Schiffer, Michael B. 1992. *Technological Perspectives on Behavioral Change*. Tucson: University of Arizona Press.

Shimada, Izumi. 1994. "Introducción." In *Tecnología y Organización de la Producción de Cerámica Prehispánica en los Andes*, by Izumi Shimada, 13–31. Lima: Pontificia Universidad Católica del Perú.

Smith, A. Ledyard, and Alfred V. Kidder. 1951. *Excavations at Nebaj, Guatemala*. Washington, DC: Carnegie Institution of Washington.

Sugiyama, Saburo. 1992. "Rulership, Warfare, and Human Sacrifice at the Ciudadela: An Iconography Study of the Feathered Serpent Representation." In *Art, Ideology, and the City of Teotihuacan*, ed. Janet Berlo, 205–230. Washington, DC: Dumbarton Oaks Research Library and Collection.

Taube, Karl A. 1992. "The Iconography of Mirrors at Teotihuacan." In *Art, Ideology, and the City of Teotihuacan*, ed. Janet Berlo, 169–204. Washington, DC: Dumbarton Oaks Research Library and Collection.

Tringham, Ruth. 1978. "Experimentation, Ethnoarchaeology, and the Leapfrog in Archaeological Methodology." In *Explorations in Ethnoarchaeology*, ed. Richard A. Gould, 169–199. Albuquerque: University of New Mexico Press.

Turner, Margaret H. 1992. "Style in Lapidary Technology: Identifying the Teotihuacan Lapidary Industry." In *Art, Ideology, and the City of Teotihuacan*, ed. Janet Berlo, 89–112. Washington, DC: Dumbarton Oaks Research Library and Collection.

Velázquez Castro, Adrián. 2007. *La producción especializada de los objetos de concha del Templo Mayor de Tenochtitlan*. México: INAH.

Widmer, Randolph J. 2009. "Elite Household Multicrafting Specialization at 9N8, Patio H, Copan." In *Houseworking: Craft Production and Domestic Economy in Ancient Mesoamerica*, ed. Kenneth G. Hirth, 174–204. New Jersey: American Anthropological Association. http://dx.doi.org/10.1111/j.1551-8248.2009.01020.x.

Wobst, H. Martin. 1977. "Stylistic Behavior and Information Exchange." In *For the Director: Research Essay in Honor of James B. Griffin*, edited by Charles E. Cleland, 317–342. Ann Arbor: University of Michigan.

Woodward, Arthur. 1941. "Hohokam Mosaic Mirrors." *Los Angeles County Museum Quarterly* 1: 6–11.

Zamora, Fabián Marcelo. 2002a. "La industria de la pirita en el sitio de Aguateca durante el periodo Clásico Tardío." Bachelor's thesis, Universidad del Valle de Guatemala.

Zamora, Fabián Marcelo. 2002b. "La industria de la pirita en el sitio Clásico Tardío de Aguateca." In *XV Simposio de Investigaciones Arqueológicas en Guatemala, 2001*, ed. J. P. Laporte, H. Escobedo, and B. Arroyo, 695–708. Guatemala: Museo Nacional de Arqueología y Etnología.

4

Whatever their function, these brilliant shining discs seem to have formed part of the ceremonial costume and must greatly have added to its barbaric splendor.

KIDDER ET AL. 1946: 130–131

Domestic Production of Pyrite Mirrors at Cancuén, Guatemala

BRIGITTE KOVACEVICH

This chapter details the evidence for pyrite production at Cancuén, which most often includes reflective mirrors (ceremonial in nature), as well as beads and dental inlays. The properties of the mineral iron pyrite are discussed, as are possible sources for the mineral in the Maya region. The social and ritual significance of pyrite and other magnetic mineral mirrors was great, and the contexts of these artifacts during the Classic period were primarily mortuary and ceremonial, although mirrors may have also been a part of domestic ritual. The use of pyrite mirrors may have been restricted by social prescriptions. It does appear that residents of non-elite structures may have been producing mirrors for elite and noble residents, as raw nodules and other production debris were found only in more humble structures. Non-elite residents may have used and possessed pyrite mirrors in domestic rituals, but production for elites almost certainly occurred and seems to have been the primary goal of household production.

As with jade production at Cancuén (Kovacevich 2006, 2007, 2011), it seems that the use of pyrite mirrors may have been restricted to certain members of the society, while others were producing them, probably for

DOI: 10.5876/9781607324089.c004

tribute. Production of these mirrors was certainly stimulated by the need for ritual paraphernalia and formed a ritual mode of production (e.g., Wells 2006; Wells and Davis-Salazar 2007). Elites were likely able to control production through ideological and esoteric means, while the products also reinforced their prestige and status through ritual communication with the gods and ancestors. At the same time, the producers of pyrite mirrors also had some prestigious items within the artifact assemblages of the household and were not simply exploited producers.

WHAT IS PYRITE?

Iron pyrite, also known as "fool's gold," can have a brilliant reflective surface like gold in a dry environment or when polished, but its iron content causes it to oxidize, turn red, and decompose with time into another mineral state in wet environments, as did the examples from Cancuén. The name *pyrite* is related to the Greek word for "fire," as pyrite will make sparks when struck with another rock (see also Zamora 2002: 29). Another good field test for pyrite is that of streak on porcelain, which is greenish-black for this particular mineral. Raw pyrite is often recovered in crystals with cubic form in numerous geological environments (figure 4.1). Pyrite is often referred to in ethnohistoric sources as "mirror stone" (see Sahagún 1950–1982, especially Book 11), but is often not distinguished from other metallic minerals that may have been used for mirror production like magnetite (Fe_3O_4), ilmenite ($FeTiO_3$), hematite (Fe_2O_3), or even obsidian. Pyrite itself in Nahuatl could be referred to as *temetztlalli* or *hapetztli* (Sahagún 1963: chapter 10).

The most common use for pyrite in ancient Mesoamerica was for mosaic mirrors. Pyrite has received relatively little attention in the archaeological literature. One major exception is Marcelo Zamora's (2002) in-depth study of pyrite use among the Maya, in which he presents evidence for the ritual significance of pyrite, contexts for its use, experimental replication studies, possible sources, and production sequences for the pyrite recovered at the site of Aguateca, Guatemala. Healy and Blainey (2011) have also recently published a comprehensive study of iron-ore mirrors among the ancient Maya.

SOURCES OF PYRITE

Pyrite is a mineral that forms as crystals within rock. This parent rock can be sedimentary, igneous, and metamorphic, making the formation of pyrite possible in almost any environment. It can occur in isolated crystals, but also in

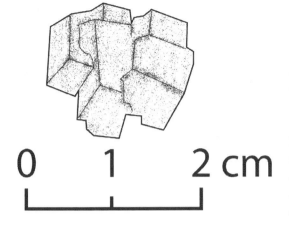

0   1   2 cm

FIGURE 4.1. *Raw pyrite nodules recovered from Structure K6-34 (drawing by Whitney Goodwin after Kovacevich 2006: Figure 6.2).*

large outcrops that can be mined or quarried. But the existence of large mines is very rare, especially in Guatemala, but more possible in Mexico (Zamora 2002: 30–31) (see chapter 2, this volume). Sahagún describes the Postclassic process of extracting pyrite or possibly other related mirror stones:

> Mirror Stones. Its name comes from nowhere. This can be excavated in mines; it can be broken off. Of these mirror stones, one is white, one black. The white one—this is a good one to look into: the mirror, the clear, transparent one. They named it the mirror of the noblemen, the mirror of the ruler. (Sahagún 1963: 228)

The raw pyrite found at Cancuén is generally cubic in form (see figure 4.1), and appears to have been collected in isolated crystals instead of cut or quarried from a mine or outcrop. During excavations, crystals of pyrite have been found embedded in limestone used in construction of residential structures. Pyrite could also have been found in the caves surrounding the Cancuén region. This was probably the most reliable source for exploitation of pyrite in the region, although the metamorphic environment where jade was quarried would also provide an area where pyrite would form and could have been collected while searching for other highland resources (see Blainey 2007: 168–177 for a thorough discussion of known Mesoamerican sources).

In his description of merchant traders, Sahagún lists pyrite mosaics (along with other fine jades, shells, and feathers) as among the objects that rulers of outlying provinces would typically send to the Aztec emperors, as tribute:

> And when the merchants reached Anauac Xicalanco [and] the rulers who governed the cities of Anauac, thereupon they gave to each of them all the

items of trade—the precious capes, precious skirts, precious shifts, the property of Auitzotzin, with which they greeted them. And then the rulers of Anauac, Xicalanco, Cimatlan, [and] Coatzaqualco reciprocated with the large green stones, round green, like tomatoes; the cylindrical green stones; then the green stones cut on a bias; the well-colored precious green stone which today we call the finest emerald green jade; and fire bottle-green jadeite, and turquoise mosaic shields; and [stones] with green pyrites in their midst . . . All this which the merchants [and] vanguard merchants took there in Xicalanco [and] carried away belonged to Auitzotzin. (Sahagún 1959: 18–19)

Zamora (2002: 31) notes that large mines or sources of pyrite are not common in Guatemala, but the archives of the Ministry of Energy and Mines of Guatemala do mention several sources for pyrite in Guatemala. In one report on geological excavations in Cerro Montecristo, Chiquimula, the presence of pyrite crystals is noted, but not the size or relative amount (MINORSA 1977). Another report (Kesler and Ascarrunz 1982) mentions pyrite outcrops in Huehuetenango, Chinuatla, San Sebastian, and San Miguel Acatán, as well as Mataquesquintla in Jalapa. Certainly not all sources or outcrops of pyrite in Guatemala have been identified, as it does not have a specific commercial value in modern society, and it is usually noted in reports investigating other minerals or rocks.

THE SOCIAL AND RITUAL SIGNIFICANCE OF PYRITE

The uses of mirrors by the ancient Maya and other peoples of Mesoamerica have been well documented (Blainey 2007; Carlson 1981, 1993; Ekholm 1945, 1973; Flannery 1968; Gullberg 1959; Healy and Blainey 2011; Heizer and Gullberg 1981; Olson 1984; Pires-Ferreira 1976; Mata Amado 2003; Nordenskiöld 1926; Taube 1992a; Merwin and Vaillant 1932: 87; Zamora 2002). The highly polished surface of mosaic mirrors not only reflected the sun, which was closely intertwined with the concept of rulership (Robertson 1974; Schele 1976), but could also possibly be used to make smoke or fire during rituals (see Carlson 1981; Ekholm 1973).

Acting as cosmic portals or devices of divination, the king or shaman could use a mirror to conjure up gods and bring them into the human realm (Carlson 1981, 1993; Miller and Taube 1993; Taube 1992a). The modern Huichol consider mirrors to be caves or passageways for gods and ancestors (see Taube 2001). This is often supported by the depiction of heads or individuals emerging from mirrors, for example at the Temple of Quetzalcoatl at Teotihuacan and

Caracol Stela 5 (Taube 1992a: 195–196, Figures 21 and 22). The use of mirrors as divinatory devices is well known around the world, especially in ancient China, for practices such as scrying (the act of seeing or telling the future with the use of a shiny object such as a crystal ball). Catoptromancy is another species of divination, performed by letting a mirror down into water for a sick person to regard his face in. If his countenance appears distorted and ghastly, it is an ill omen; if fresh and healthy, it is favorable (Webster's Revised Unabridged Dictionary 1998; e.g., Besterman 1997).

Taube (2001) argues that the relation of mirrors to caves and portals can also be seen during the Late Classic and the Postclassic periods, with their representation being that of the "middle place" or the symbolic opening between the earth and the heavens and underworld. He argues that mirrors were sometimes worn on the abdomens of figures to represent the earth navel or the middle of the cosmos.

Among modern Maya, the word *lem* in Eastern Maya languages, and *nen* (Schele and J. Miller 1983: 10–12) or more recently *nehn* (Stone and Zender 2011: 73) in Western and Central Maya languages is "mirror," or the quality of reflecting or illuminating (see also Healy and Blainey 2011). Stuart (2010: 291) reads the mirror glyph (T617) originally identified by Jeffrey Miller as *lem* as generally meaning something shiny or resplendent (he calls it "shiner"), which could extend to various types of shiny stones, such as jade or obsidian. Healy and Blainey (2011: 235) argue that the T617 and T24 glyphs, while associated many times with shiny stones or surfaces in general, are good candidates for representations of reflective mirrors (see also Macri and Looper 2003: 274–275), especially given that T24 can have the reading *li-* or *-il* and can be associated with the morpheme *il*, "to see" (Houston, Stuart, and Taube 2006: 139), connecting eyes and mirrors as "liminal thresholds," and also reinforcing the role of mirrors in shamanistic endeavors (see also Blainey, chapter 9, this volume).

Mirrors could also act as a representation of kingship. Schele and Miller (1983: 12) report that there is an important description of the use of the word *nen* related to titles for priests and holders of office in the Motul Dictionary: "*u nen cab, u nen cah* el sacerdote, cacique, gobernador de la tierra o pueblo, que es el espejo en que todos se miran," or "the mirror in which all people see themselves reflected" or the people's or the earth's mirror (see also Carlson 1981, 1993). The ruler as the people's mirror can also be seen in the following quote from Nezahualpilli as Montezuma takes the throne:

O you most powerful of all the kings of earth! The clouds have been dispelled and the darkness in which we lived has fled. The Sun has appeared and the light of dawn shines upon us after the darkness which had been caused by the death of the king. The torch which is to illuminate Mexico has been lighted and today we have been given a mirror to look into. (Durán 1994: 220)

Some scholars also believe that there was a mirror ceremony, which marked heir designation or ascension to kingship, and in Maya iconography the grasping of a mirror is often indicative of the ascension of a new ruler (Schele and J. Miller 1983). Schele and J. Miller (1983: 14) note that the term *nen* denotes rulers and persons of importance, "the reflection of the world (or people)" in Yucatec. In Quiche, *nem* can mean "succession in office." This reading for T617a (the *nen* mirror glyph) for the Classic Maya is supported by its use in iconographic contexts of accession and heir-designation, especially when it is followed by a prepositional phrase designating the person "as Enterer of the Succession." Schele and Miller argue for textual evidence of this at Palenque. While new advances in epigraphy and iconography may call some of these readings into question, Healy and Blainey (2011: 236) argue that the incorporation of T24/T617 in glyphic representations of ascension demonstrates the king's role as mediator between different worlds.

Matsumoto (2013) suggests that certain Maya monuments that were written in mirror image to commemorate political achievements were also evoking the power of the mirror itself and its connection to political and ritual power. Stuart (2010) makes the argument that many monuments were polished to create shiny surfaces to mimic other shiny stones, like jade celts worn on the belts of rulers that recall the power of Chahk and flashes of lightening. Houston, Stuart, and Taube (2013: 67) also detail the importance of polished stone stelae with depictions of rulers impersonating gods, which they interpret as having played an important role in rituals for the ascension of a new king (again, connecting very polished or shiny stones to ascension and kingship).

The *Chilam Balam of Chumayel* (Roys 1933: 109–110) also chronicles the use of the mirror metaphor in birth and ascension:

> In the first katun was born the only son of God; in the second katun, the Father. In the third Katun was *Expleo-u-caan*, as he was called, who chastised him named *Chac Opilla* when he set up the heavens. *Enpileo-u-caan* was his name. *Expeo* was his name within the first noose of God. Hebones was the only son of God. [Like] a mirror he was born astride on the shoulder of his father, on the stone of his father. (Quoted in Schele and J. Miller 1983: 19.)

Thus as the god is born into divine status he becomes like a mirror, just as a king becomes a mirror as he ascends to the throne.

Mirrors were obviously associated with the divine, but there are also clues as to how these objects were worn by those who were impersonating gods. Forehead mirrors and other body mirrors were often associated with the Sun God, the so-called Jester God, God K (also *Kawil* or *K'awiil*, see also Stuart 2010), God D, and variants of God C (Miller and Taube 1993; Schele 1974, 1976; Schele and J. Miller 1983; Taube 1992b; see Schellhas 1904 for original designation of lettered gods, and Taube 1992b: 148 for further explanation). Schele and J. Miller (1983) argue that God K is often represented with the dark (possibly obsidian) mirror, while the Jester God is represented with the light mirror (probably pyrite or other magnetic minerals). Thus the ruler would become the personification of these gods as he wore the mirror:

> The "mirror" grapheme has been shown to appear as a major feature in the diagnostic traits of God K and the Jester God (both major gods of Classic rulership) as well as other deities. The prominence of both of these deities in the regalia and accoutrements of royal portraiture is relevant. If both gods participate in royal iconography and mythology, it is little wonder that the forehead motif of each becomes a symbol of accession to office in general and to the throne in particular. (Schele and J. Miller 1983: 14)

The Jester God, so named because his tripartite headdress looks like that of a medieval European jester, has been tied by many scholars to the personification of kingship (e.g., Schele and M. E. Miller 1986: 79), but has been more recently interpreted to have at least three distinct forms by Stuart (2010) and Taube and Ishihara-Brito (2013: 147). It is the "avian" form of the god (Taube and Ishihara-Brito 2013: 149–150) that is sometimes shown with a mirror element (see Taube and Ishihara-Brito 2013: Figure 81c, and compare Schele and J. Miller 1983: Figure 3g). Stuart (2010) argues that this form of the deity is associated with the "spirit" of *amate* bark paper, which was also intimately tied to inauguration and crowning of Maya kings, often incorporated as a paper headband that was "fastened" to the new king.

It is the piscine form of the Jester God that is often worn on the forehead of a ruler as a headdress ornament in jade (or alabaster at Aguateca [Triadan 2000; Eberl and Inomata 2001]; see also Taube and Ishihara-Brito 2013: 148). The piscine Jester God headdress recovered from Aguateca also had numerous pyrite incrustations that surrounded the headband, possibly recycled from pyrite mirrors, which is also seen in the iconography representing Ruler 5 on Stela 19 at Aguateca (Zamora 2002: 13). This indicates that in at least one case the piscine form of the Jester God was associated with pyrite.

God K is associated with lightning, rain, maize, lineage, and rulership (Coggins 1975, 1979, 1988; Taube 1992b). Taube (1992b: 79) also argues that the agricultural side of God K is intricately linked to elite power and dynastic descent. God K's association with mirrors is also supported by the four wooden figures recovered at Tikal depicting the god holding a mirror plaque in front (W. Coe 1967: 57; Taube 1992b: Figure 36), which Taube notes is very similar to the back mirror of the effigy from Burial 10. Taube (1992b: 76) also argues that the mirror representation of God K, as well as related elements of fire, burning axes, torches, and cigars, may all refer to lightning, and this may be a reason that God K is often counterpoised with Chahk. The association of mirrors and God K with divination is also important. Zamora (2002) identifies the connection between God K and the Postclassic god Tezcatlipoca (Taube 1992b) and points out that Sahagún (1985: 195) mentions a divinatory role for that god.

Ethnohistoric sources also associate mirrors with rulers, gods, and rulers imitating gods, as seen in the burial of the Aztec king Tizoc:

> After them came the King and Lord of the Underworld, dressed like a diabolical creature. In the place of eyes he wore shining mirrors; his mouth was huge and fierce; his hair was curled; he had two hideous horns; and on each shoulder he wore a mask with mirror eyes. On each elbow there was one of these faces, on his abdomen another, and on his knees still other faces with eyes. With the shining of the mirrors that represented eyes on all these parts, it looked as if he could see in every direction. (Durán 1994: 308; a footnote says he must be the god Tlatecuhtli)

There are also references by Landa to mirrors being the sole property of men, and also representing cuckholdery: "All the men used mirrors while the women had none; and to call each other cuckholds, they said that the wife had put the mirrors in the hair at the back of their heads" (Tozzer 1941: 89). As Zamora (2002:16) suggests (see also Schele and M. E. Miller 1983: 12), this probably means that the usual position for mirrors would be to be worn on the forehead or on the front of the head. Mirrors appear in headdresses depicted on murals at Teotihuacan (Miller 1973: Figures 202, 210, 211; Langley 1986: Figure 32), as well as on a Teotihuacan-style stela from Kaminaljuyu (Parsons 1986: Figure 190).

The use of mirrors as pectorals in Mesoamerica is also well documented in iconography and burial contexts, such as figurines at Teotihuacan (von Winning 1987), in tombs at Kaminaljuyu (Kidder et al. 1946: 126), as well as on Olmecoid figurines (Taube 1992a: 178). Mirrors in Mesoamerica were also often worn on the small of the back, and in the Postclassic were called a *tezcacuitlapilli*. Moreover, mirrors had Early Classic counterparts, especially at Teotihuacan where they can be seen in the iconography, especially in warrior costumes (Taube 1992a: 173), and in dedicatory burials in the Temple of Quetzalcoatl (Sugiyama 1989: 97) and the Pyramid of the Moon (Sugiyama and López Luján 2007). Back mirrors were associated with two burials in Tomb B-I at Kaminaljuyu during the Esperanza phase (Kidder et al. 1946). These can be seen on the Postclassic Atlantean figures at Tula and have been found in archaeological contexts in the Temple of the Chac Mool at Chichen Itza (Morris et al. 1931). Taube (1992a: 198) argues that back mirrors were more of an Early Classic innovation, whereas Olmec mirrors were primarily worn on the chest or forehead and were of a smaller size. This could possibly be due to the use of pyrite among the Early Classic Maya (instead of magnetite or hematite as with the Olmec), which allowed the elaboration of larger mirrors.

The association of mirrors with war, warriors, and warfare is present across Mesoamerica. Taube (1992a: 192) notes that they could have in some cases served protective or possibly spiritual functions. He notes that Tezcatlipoca, the smoking mirror god, was seen as a warrior in the Postclassic. Mirrors represented aspects of fire and water, also associated with the Aztec concept of war *atl-tlachinolli*, or water-fire (Séjourné 1957). The mirror is also often seen as a representation of the sun. In Classic Maya art mirrors are represented with the *kin* sign, as well as Aztec representations of the smoking fifth sun (Taube 1992a: 193–195). The modern Huichol people of Mexico also see the sun as representing a mirror (Carlson 1981: 125).

The depiction of mirrors in Classic Maya ceramics also supports not only the connection of mirrors to the royalty and nobility, but also a divinatory rather than a banal function for the mirrors themselves (see Rivera Dorado 1999). Mirrors are often pictured in royal courts as evidenced in ceramic iconography. These mirrors were often illustrated as quite large, which may have meant that they were pyrite mirrors; as Taube (1992a: 179) points out, mirrors of pyrite were the only ones with virtually no size limit, as they were made of mosaics, whereas mirrors of obsidian or other metallic minerals would be limited to the size of the original nodule. One example is cylindrical vase MSO488 (see Reents-Budet 1994: 83, 92, 322), which depicts three Maya elites involved in a divinatory ritual: the central figure, adorned with jade, stares into a large mirror held by an attendant. A bundle can be seen nearby, possibly holding ritual paraphernalia. Another attendant sips from a cup, possibly holding a ritual or hallucinogenic substance (see Blainey, chapter 9, this volume).

Miller and Taube (1993: 79) note the importance of the connection between the acts of divination and creation, as the Divine Ancestral Couple and the creators themselves are characterized as diviners. In Central Mexico, Oxomoco and Cipactonal were considered the primordial couple as well as diviners, and the similar couple in the Quiche Maya *Popul Vuh* (Xpiyacoc and Xmucane) performed a divinatory hand casting of sacred stones during the creation of humans. The Aztec goddess Toci was seen as a diviner as well as mother of the gods, similar to the Maya Goddess O, associated with creation and destruction, childbirth, and especially divination. Taube (1992a: 103) finds that Goddess O was depicted in the Dresden Codex page 42a sitting upon a pyramidal structure holding a mirror bowl with the face of God C. Thus, the acts of creation and divination were inextricably linked throughout Mesoamerica, and the mirror was an important ritual tool in those acts. For example, among the Tarascans of Michoacan, the shamans of the king could see all past and future events through bowls of water or mirrors. The events witnessed by these seers could be used as evidence in court cases (Miller and Taube 1993: 80).

Taube (1992b: 181) finds that in different contexts mirrors at Teotihuacan can represent human eyes, faces, flowers, fiery hearths, pools, webs, shields, the world, the sun, and caves or passageways. It is not difficult to see how the relationship of mirrors and eyes can be conflated in the notion of divinatory practices, as "seeing" itself can take on both meanings. In the Florentine Codex, eye, pupil, and mirror, are all represented by the word *tezcatl* (Taube 1992b).

Paragraph 5 of chapter 8 of Sahagún's (1963) chronicle of *Earthly Things* in Book ii details the colonial account of "mirror stones":

> Fifth Paragraph, which telleth of still another kind of stone. It is a stone [of which] mirrors are made, [a stone] which is converted [into a mirror] . . .
>
> The black one—this one is not good. It is not good to look into; it does not make one appear good. It is one (so they say) which contends with one's face. When someone uses such a mirror, from it is to be seen a distorted mouth, swollen eyelids, thick lips, a large mouth. They say it is an ugly mirror, a mirror which contends with one's face.
>
> Of these mirrors, one is round; one is long; they call it *acaltezcatl*. [These mirror stones] can be excavated in mines, can be polished, can be worked.
>
> I make a mirror. I work it. I shatter it. I form it. I grind it. I polish it with sand. I work it with fine abrasive sand. I apply to it a glue of bat excrement. I prepare it. I polish it with fine cane. I make it shiny. I regard myself in the mirror. I appear from there in my looking mirror; from it I admire myself.

Mirrors were a symbol of kingship, divinity, and ritual power. In the context of Classic Maya ideology, these symbols were an authoritative resource, mobilized by the elite to reinforce and legitimate their power and special position within the social structure. Again pyrite was a symbol of status and power, but also an actual tool used for conjuring, divination, and production of smoke (and possibly fire) for ritual purposes. It is clearly stated in the passage above: "They named it the mirror of the noblemen, the mirror of the ruler." The use of mirrors defined the ritual practitioner as a living god, endowed with the divine power to communicate with the Otherworld and to see the future. Acts of divination were also likened to acts of creation, again reinforcing the identity of the elite user as a creator of humanity, one of the few individuals with the power to recreate the cosmos on a daily basis, as well as further the sacred covenant with the gods and ancestors.

PYRITE CONTEXTS IN THE CLASSIC MAYA WORLD

Much of the information on contexts for pyrite mirrors comes from the work of Marcelo Zamora (2002; see also Blainey 2007: Appendix A; Healy and Blainey 2011); the following discussion will add some additional information to his comprehensive list. I report here only those examples for which there is an explicit provenience, as context is the most important factor in this analysis. My main objective in this section is to identify an "elite" versus "non-elite" context for other Maya sites in order to compare these to Cancuén.

From Zamora's (2002: 31–41) summaries we can determine that pyrite mirrors or their tesserae (tiles) were found in elite caches, tombs, or structures at Pusilha (Joyce, Clark, and Thompson 1927: 449), Piedras Negras (Barrientos et al. 1997; W. Coe 1959: 42–43; Escobedo and Alvarado 1998; Escobedo and Zamora 1999; Houston and Arredondo 1999: 253; Houston and Escobedo 1997, 1998, 1999), Aguateca (Inomata and Triadan 2000; Zamora 2002), Copan (Davis-Salazar and Bell 1999: 1113–1128; Sanders 1990), Chichen Itza (Morris et al. 1931; Ruppert 1935: 36), Dzibilchaltun (Taschek 1994: 97–99), Mayapan (Proskouriakoff 1962: 354), San Jose (Thompson 1939: 20, 47, 176, 184, 188–189), Holmul (Merwin and Vaillant 1932: 87), Uaxactun (Kidder 1947: 52), Topoxte (Fialko 2000: 144–149), Altun Ha (Taschek 1994: 97–99), Chama (Shook and Kidder 1952), Zacualpa (Wauchope 1975), Nebaj (Smith and Kidder 1951: 87, 102), Zaculeu (Woodbury and Trick 1953: 87, 102, 115), and Kaminaljuyu (Kidder et al. 1946; Shook and Kidder 1952: 126–134).

Some additional elite contexts I found in the literature that contain pyrite (or other mosaic) mirrors include an Early Classic cache at Quirigua (Ashmore 1980, cited in Taube 1992b: 177). Tomb 116 of Tikal in Temple 1 contained 3 mosaic mirrors (Harrison 1999). Mosaic mirror pieces were also recovered at Machaquila from Structure 4, an elite structure in the epicenter of the site (Ciudad Ruiz et al. 2003: 250). A mosaic blue hematite crystal and mother-of-pearl mirror was discovered in a royal cache in Structure 6B at Cerros, along with the royal jade jewels of the first king (Schele and Freidel 1990: 121). Excavations at Thompson's Group, Belize revealed a cache in the principal group, which contained numerous exotic items, including pyrite (Kunen 2004: 82,102; Robichaux 1995: 390–395). Recent excavations at Pusilha, Belize, by Braswell and colleagues (2005) have also uncovered a pyrite mirror and hematite sequins associated with Burial 6/1, a secondary elite interment located in a pyramidal structure in the epicenter of the site. Also recovered from the Pusilha tomb of a possible ruler (Burial 8/3) were a total of 81 jade artifacts, including 24 beads and 3 carved jade ornaments, a large *Spondylus* shell, and a fragment of pyrite, among other goods (Braswell et al. 2005: 79–80). A pyrite mirror was recovered from the Margarita tomb in the Copan Acropolis (Bell 2002: 99). In the Sepulturas elite residential zone of Copán, four slate or mudstone mirror backs were recovered, one highly polished concave stone, and two (possibly three) pyrite mirror pieces (Willey et al. 1994: 251).

Importantly, some "mirror stones" have been recovered from non-elite or lesser elite contexts (see Gonlin 2007); for example, the Rio Amarillo area, a more humble residential area 21 km from the Main Group at Copan. Structure 1 (Site 7D-3-1), where the cache was recovered, was not a large, vaulted mound, but one with a stone cobble base and perishable superstructure. This cache contained "a fishhook-shaped obsidian eccentric on the surface; and from the base of the cist, two complete ceramic vessels, a 10 cm long chert spear point, and a small extremely thin piece of jade" (Gonlin 2007: 97). The modest-sized site also yielded a cache with a "small, round mirror polished from some iron-bearing mineral—one of few such mirrors ever found at Copán" (Webster and Gonlin 1988: 178). Burial 2 in the B Group, a pair of small structures at the Bajo Hill site of Belize contained three small, polished hematite disks (the largest approximately 2 cm in diameter). This group is about 40 m from the largest group at the site, and the structure itself had a limestone foundation brace wall, with a plaster interior floor (Kunen 2004: 64). A rural sweatbath at the site of Piedras Negras also contained a pyrite mirror (Child and Child 1999), reaffirming the role of mirrors in shamanic ritual. These findings suggest that pyrite or other mirror stones may have been the property of non-elites in some cases. As archaeological studies increasingly focus on household archaeology, we may find an increase of polished iron-ore items in these types of contexts.

TABLE 4.1. Mean length, width, thickness, and weight for finished mosaics

Pyrite mosaic size	Mean length	Mean width	Mean thickness	Mean weight
Non-masonry structures	19.14	15.41	2.78	2.79
Masonry structures	26.32	17.92	3.00	6.25

PYRITE MANUFACTURING SEQUENCE

The study of the manufacturing sequence of pyrite can again be informed by ethnohistoric sources, as well as replication experiments. Many of the technological steps and tools used are very similar to that for jade (see Kovacevich 2011), suggesting that those who produced goods in one medium could have produced them in the other. This seems to be the case in the archaeological record at Cancuén. The tools and manufacturing steps were recorded and compared for different contexts at Cancuén and are discussed below. Special attention was paid to the stages of production, distributions of raw materials, half-finished artifacts, and tools used in production. The stages of production for Cancuén are interpreted from indirect evidence of tools associated with production contexts and ethnohistoric sources, as no replication experiments were performed for this material (see Gallaga, chapter 2, this volume; Melgar et al., chapter 3, this volume; and Zamora 2002, for results and discussion of experimental studies concerning pyrite). The significance of the present study lies in the context of production refuse and finished products, which in many cases at the site is in non-elite residences.

Raw pyrite was probably shaped into mosaic mirror pieces by stone grinding tools used with an abrasive such as quartz. Then the mirror was polished with substances like hematite, which is essentially the same as jewelers' rouge used today. Pyrite could also be formed into beads and dental inlays, but it was not used for weapons or tools. In other words, while pyrite is a metallic mineral, its processing and production into artifacts in no way involved smelting or the technology of metallurgy, which fully developed as a technology during the Postclassic period (Bruhns 1989).

The first step in working pyrite after the procurement of the raw material would be either percussion, sawing, or shaping, depending on the size of the nodule and the desired end product. Zamora (2002: 22) notes that the use of percussion (indirect in his reproduction studies) produces flakes with very irregular surfaces, which would have required an extensive amount of work in shaping and perfecting before they could be polished (see also Gallaga, chapter 2, this volume). Zamora (2002: 23) also notes that cutting or sawing would produce perfect preforms for mosaic pieces, and some samples of mosaics from Aguateca do demonstrate evidence of sawing. This notion is supported by the presence of string-saw anchors in association with pyrite production areas at Cancuén, and the process was surely similar to that used for jade and other lapidary work (see Kovacevich 2011).

Toscano (1970) discusses Aztec techniques for the creation of mosaics recounted by Sahagún: first the mosaic pattern would be drawn on paper, then the pieces would be cut and shaped by the lapidary to fit the pattern. As Toscano (1970: 175) explains:

> Las láminas de piedras finas eran raídas con cuchillos y después bruñidas con esmeril, y finalmente montadas y pegadas con goma tzauhtli, y las raen con un pedernal partido; y las ahuecan y hornadan con un tubito de cobre. Después les hacen facetas muy cuidadosamente, las bruñen y les dan lustre. Las pulen montadas en madera, de suerte que se ponen muy brillantes, radiosas, lucientes.
>
> [The tesserae of precious stones were scratched with knives and then polished with emery, and finally assembled and held together with *tzauhtli* adhesive (glue extracted from orchid leaves), and then they incise them with a split flint, and then they perforate them and create depressions with a copper tube. After that they make facets very carefully, and burnish them and give them luster. They polish them mounted in wood, so that they get very bright, radiant, shining.] (Translation by the author.)

There is little mention of the direct process of sawing in the ethnohistoric literature, but the string-saw is a possible option as that would produce the thin flat preforms, whereas wedge saws or stone saws would leave a more inverted surface that would not be as flat. Replication work by Melgar et al. (chapter 3, this volume) suggests that obsidian blades were also important in this stage.

Shaping or abrading would be another option for creating mosaic or bead preforms. Some pyrite artifacts at Cancuén show signs of having been abraded, possibly with a grinding stone, although no definite tools with this quality were found directly associated with the objects. This would have been a laborious process, as pyrite is a very hard mineral itself, 6–6.5 on the Mohs scale of hardness. This is probably also how pyrite beads were shaped. Abrasion and sawing could have been aided with fine sands or emery.

Sahagún (1963: 238) describes "flint sand," possibly referring to ground chert:

> It is crushed, pulverized, ground. It is flint. It is fragmented, broken up, fine, much ground, very fine, completely ground, all ground like *pinole*; powdery. It is

a medium for cleaning, for polishing, for thinning, for scouring. It is a polisher of things, a smoother of things. I polish something. I smooth something. I dress something.

The use of an abrasive such as this for shaping is understandable as it would have a hardness of 7, greater than that of pyrite. In describing the activities of the lapidary, Sahagún (1963: 238) refers to abrasion with *teuxalli,* or emery (also called corundum, with a hardness of 8–9):

It is now called *uei xullli.* [It is of] the mountains, the crags. It is white, yellow, ashen, ruddy, mixed black and green. It is a grinder; it is that which wears away, which thins things. Some are like volcanic rock, rough, fragmented; some are fine, quite minute. They grind, abrade, thin, wear things away. I treat things with *teoxalli.* I abrade things. I harden.

After cutting and shaping, the pieces must be polished to create the shine and brilliance associated with the finished product. All of the pyrite artifacts recovered from Cancuén are now oxidized and do not retain their original luster, but surely were polished as Sahagún recounts for Aztec lapidaries:

The Mirror-Stone Seller, the mirror-stone maker [is] a lapidary, a polisher. He abrades (with the instrument called *teuxalli*); he uses abrasive sand; he cuts; he carves; he uses glue of bat excrement, polishes with a fine cane, makes it shiny. He sells mirror-stones—round, circular; pierced on both sides, [which] can be seen through; two faced, single faced, concave; good mirror-stones, white mirror stones, black mirror stones. (Sahagún 1981: 87)

Polishing was most likely accomplished with cane (mentioned above), soft stone powder (like pumice, calcite, or finely ground emery; see Zamora 2002: 25), or possibly leather (Melgar et al., chapter 3, this volume).

If the end product was a pyrite mirror mosaic, the pieces would have been assembled on a mirror back, usually made of ceramic or slate, but possibly also made of wood, especially for the larger mirrors pictured in ceramic scenes of the royal court (although preservation in the Maya world has rarely permitted the recovery of a wooden mirror back, see Blainey 2007: Figure 21 for an example from Chichen Itza). These backs nearly always have holes drilled in them so that they may be hung as an adornment (usually on the head, chest, belt, or back, as discussed above). Kidder et al. (1946: 126) argue that there is a chronological component to the type of holes drilled in the mirror backs, whether they were holes drilled straight through or with a channel allowing a string to set into the mirror back. It has not yet been determined exactly how the mosaics were affixed to the base. Sahagún's statement above mentions bat excrement, whereas specimens at Aguateca (Zamora 2002: 25) and Cancuén (Kovacevich 2006: 249) retain evidence of stucco. It is possible that other types of natural resin were used as adhesives (see Gallaga, chapter 2, this volume).

PYRITE PRODUCTION AT CANCUÉN

Evidence of pyrite mirror production in non-elite structures at Cancuén includes raw pyrite nodules, half-finished mosaic pieces, broken mosaic pieces, finished mosaic pieces, and stone and ceramic mirror backs. Nodules and mosaic pieces were identified as pyrite by Duncan Cook with a scanning electron microscope and X-ray diffraction at the Smithsonian Center for Materials Research and Education (SCMRE). Geochemical evidence from structure M6-12 also revealed levels of iron five times higher than normal in a localized spot on the patio/production area. The high iron levels in the floor could possibly relate to pyrite production (Cook et al. 2006; Kovacevich et al. 2004; see Inomata et al. 2002 for a similar interpretation), supported by the presence of raw pyrite nodules and finished and broken mosaics in midden and floor contexts (see figures 4.1 and 4.2).

All pyrite artifacts at Cancuén were recorded for maximum length (mean 2.25 cm, sd 1.14), width (mean 1.75 cm, sd 9.33), and thickness (mean .55 cm, sd 4.74). The total weight of pyrite artifacts from Cancuén is 11.96 kg, with a mean weight of 12.46 g and a standard deviation of 70.76. The mean thickness and weight of whole pyrite mosaic pieces found in all structures are strikingly similar (see table 4.2), with the exception of one mosaic found in an elite structure (N10-1) which was thicker, larger, and heavier than any other found at Cancuén (30.61 × 18.37 × 5.33 mm, 9 g), and was also found just below the surface. Although no mosaics or pyrite nodules were found in earthen mounds with no stone, the most pyrite in mosaic and nodule forms was found in low mounds either with low retaining walls with an earthen core or those with earthen mounds and a limestone *laja* (i.e., flagstone) exterior patio floor.

PYRITE CONTEXTS AT CANCUÉN

One hundred and one pyrite or pyrite-related objects were recovered from residential excavations at Cancuén from 1999 to 2003 (see tables 4.1, 4.2, and 4.3). The largest category of pyrite objects were pyrite mosaic pieces or fragments of those mosaics that would have formed mirrors, including 65 pieces

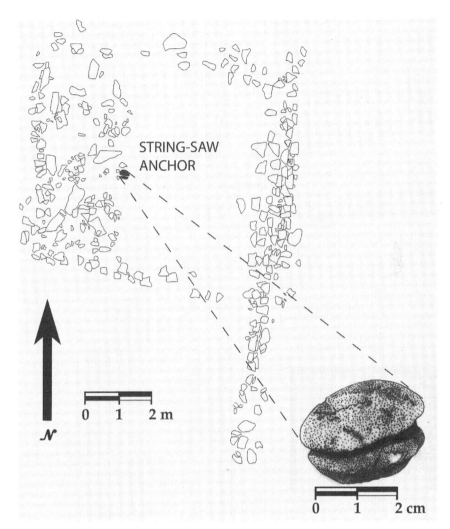

STRING-SAW
ANCHOR

0 1   2 m

N

0   1   2 cm

FIGURE 4.2. *Plan view of Structure K6-34 showing the location of the string-saw anchor (drawing by author, modified from Kovacevich 2006: Figure 6.6) and a drawing of the string-saw anchor recovered (drawing by Laurie Greene).*

in all. Five unfinished mosaics were recovered from a single structure, K6-34 (see figures 4.1, 4.3, and 4.4; Barrientos et al. 2000: 124–128; Kovacevich 2002). This evidence seems to demonstrate that the residents of at least this structure were involved in the production of pyrite mosaics. The only two limestone

TABLE 4.2. Distributions of pyrite artifacts		TABLE 4.3. Contexts of pyrite artifacts	
Pyrite artifacts	*Totals*	*Pyrite contexts*	*Totals*
Nodule	17	Humus	18
Mosaic (finished)	28	Fill	7
Mosaic (unfinished)	5	Wall fall	23
Mosaic (fragment)	32	Floor (interior)	15
Mirror back (fragment)	13	Floor (exterior)	22
Bead	2	Midden	12
Dental inlay	4	Burial	4
Total	101	Total	101

mirror backs, one complete and one fragmentary, were also recovered from this structure on the interior floor. Both of the mirror backs were perforated with two holes near one edge of the object, suggesting that they were made to be worn as adornments. Of the 15 nodules recovered from excavations at Cancuén, four were from Structure K6-34. All of the pyrite artifacts from K6-34 came from floor or wall fall contexts. Also located in the floor context was a string-saw anchor (see figure 4.2), which was a tool certainly used in the production of jade in other contexts at Cancuén (see Kovacevich 2011), but may have also been used to saw off pyrite mosaic pieces before shaping and polishing. A pyrite bead weighing 5.5 g was also recovered from the interior floor of the structure. The contexts of these artifacts primarily on interior and exterior floors suggests again, as with jade working, that these artifacts were left in a rapid abandonment of the site as de facto refuse, and also possibly represent termination rituals (cf. Garber 1993 for a similar example with jade for non-elite residences at Cerros).

The other raw nodules that were recovered were from earthen structures with limestone *laja* exterior patio floors—K7-24, M10-6, M10-7, M6-12 (see figure 4.3)—and two other earthen structures. One pyrite bead weighing 0.3 g was recovered from the exterior floor of Structure M10-4, and eight small jade beads were found embedded in the earth surrounding the patio floor stones of a floor related to jade production. The only pyrite artifacts recovered from burial contexts were four dental incrustations in Burial 66 of a masonry structure in an elite compound, N11-1 (figure 4.4; Kovacevich et al. 2003), which has also been found for elite burials at sites like Holmul (Merwin and Vaillant 1932: 33). No pyrite artifacts were recovered from caches or ceremonial deposits.

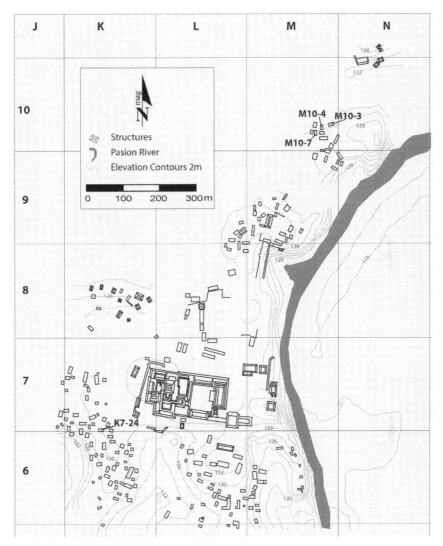

FIGURE 4.3. *Map of Cancuén (redrawn by Lia Tsesmeli after Kovacevich et al. 2005: figure 4.2).*

Two elite masonry structures, K8-1 (one mosaic) and N10-1 (three mosaics) had finished, complete mosaic pieces recovered in the humus (figure 4.4), approximately five cm below the surface, just above architecture and wall fall. These contexts again may be due to termination rituals of elites or priests

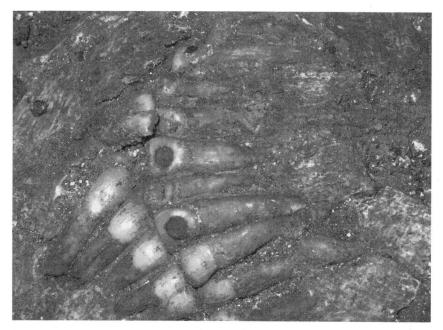

FIGURE 4.4. *Jade- and pyrite-inlaid teeth, Burial 66, Structure N11-1 (photo by author).*

that involve ritually destroying objects that hold ceremonial power before abandoning the site (Hendon 2000; see also Mills 2008 for an example from the Southwest). This is a pattern also seen with broken jade artifacts (see Kovacevich 2006). No evidence of raw nodules or mosaics in the process of production was recovered from any context in elite masonry structures.

DISCUSSION

The mean thickness of pyrite mosaic pieces (which were found in all structures) do not differ significantly, with the exception of one mosaic found in an elite structure, N11-1 (see also Kovacevich 2007). This does not necessarily suggest that there were different industries creating larger or more elaborate mirrors for elites, as the thickness of mosaics in elite structures were comparable to those from earthen mounds. No whole pyrite mosaics were recovered from the smallest mounds with the least amount of investment in labor. The most pyrite by far, in mosaic and nodule form, was found in what appear to be non-elite structures. No raw nodules were recovered from masonry structures, although loose mosaic pieces were recovered.

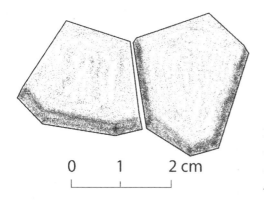

0   1   2 cm

FIGURE 4.5. *Pyrite mosaic pieces recovered from Structure N10-1 (drawing by Whitney Goodwin after Kovacevich 2006: Figure 6.12).*

This evidence contrasts with that found at Aguateca, where there was elite production of pyrite mosaics (Inomata and Triadan 2000; Zamora 2002). However, if elites were producing mirrors one would expect some production debris, abandoned mosaics in the process of production, and/or tools in middens or other contexts if elites were producing mirrors at Cancuén. No raw nodules, half-finished mosaics, or string-saw anchors were encountered in masonry structures. It should be noted that to date the only pyrite found in a burial or tomb at Cancuén was from the four dental incrustations of Burial 66 in an elite residential group, N11-1 (Kovacevich et al. 2003).

The question that follows is whether the residents of more humble structures were producing mirrors for elites. It certainly seems that non-elite residents were producing mirrors for household use, domestic ritual, and/or divination. Recent research has emphasized the overlooked aspect of non-elite domestic ritual in Maya archaeology (Gonlin and Lohse 2007; Robin 2003). A domestic use of mirrors in these structures would support an enhanced status for the residents, for, as was discussed above, the context and use of pyrite mirrors in the ancient Maya world was almost exclusively in the elite domain. This status may have resulted from the production for exchange or tribute of other important items, as in at least two cases—Structures M10-4 and M6-12—where raw pyrite and possible production was associated with jade production.

It seems that non-elites possibly produced mirrors (at least sometimes) for elites. In this study, non-elite residences were identified by low labor investment in construction, but also by the lack of hallmarks of elite culture, such as stone masonry, vaulted architecture, hieroglyphic writing, and tombs or burial cysts, which appear to have been restricted to the elite (see also Kovacevich 2006). First, the lack of pyrite production evidence in elite structures, but the presence of mosaics (albeit single mosaic pieces separated from a mirror back), suggests that they owned them but did not produce them. Second, the size, thickness, and weight of the mosaics did not differ significantly between house types (Kovacevich 2007). The presence of pyrite in an elite burial suggests that it was valued by elites; a similar finding of pyrite dental incrustations was recovered from Burial 5 of K'inich Yo'nal Ahk II of Piedras Negras (see W. Coe 1959 and Sharer and Traxler 2006:426). The fact that two mirror backs were recovered from Structure K6-34, one of fragmentary slate and one whole ceramic back, suggests production for tribute or exchange. Two mirrors present in a single small structure would seem to represent something above that needed for household use (although no real precedent exists). The contexts of these mirror backs, as well as the raw nodules and mosaics in earthen mounds also suggests production, as they often were recovered from floors and middens, rather than caches or burials. They were not carried away as treasured valuables or heirlooms as the site was abandoned.

The production of pyrite alongside that of jade would also make sense, as many of the lapidary techniques described in ethnohistoric sources for the production of mirrors were very similar to the early stages of jade production (see also Kovacevich 2011). If elites were levying tribute in the form of unfinished jade plaque or earflare blanks from these residents, it would be easy to add pyrite production to these requests, as it would require nearly identical knowledge, skills, and tools. Given that the amount of time invested in the creation of a pyrite mirror could top 100–150 person-hours (Gallaga, chapter 2, this volume), the segmentation of production could be a reasonable way for production to be more efficiently completed (see Melgar et al., chapter 3, this volume). Thus, both elites and commoners may have been involved in production of these pieces.

I am not discounting the possibility that these mirrors could have been produced for domestic use and ritual (although cf. Gonlin 2007; Webster and Gonlin 1988). Combined with evidence of jade production and use, data suggest that certain objects were restricted to elite use, and mirrors may have been one of these, especially because the ritual knowledge used to conjure, divine, and/or impersonate gods may have been the sole property of a certain elite group in the society, or even the ruler himself. Prestige and status were certainly accorded to these producers, as they themselves were part of the creative process, which itself was divine (see Hruby 2006; Reents-Budet 1998), while some functions and uses of the mirrors may have been restricted to the elite.

Although the non-elite did not live in masonry structures with high investment in construction, or were not located in the epicenter of Cancuén, their burials did have signs of higher prestige and status when compared with other households, including a warrior figurine recovered with a burial from Structure

K6-34 (Sears 2000) and Chablekal Fine Gray ceramics from Structures M10-7 and M10-4 (Bill and Callaghan 2001; Bill et al. 2003; Callaghan and Bill 2003; Callaghan et al. 2004; see also Kovacevich 2007).

CONCLUSIONS

Until recently, pyrite has received relatively little attention when compared to other "prestige" or ritual items and has been assumed by many to be an elite good. Contextual evidence across many ancient Maya sites largely supports this assumption, although recently some scholars have found that "mirror stones" or mirrors may have been used in non-elite domestic rituals as well (Gonlin 2007; Webster and Gonlin 1988). Evidence at Cancuén suggests that non-elites *could* have used mirrors in domestic ritual, but the possibility still exists that they produced the mirrors for elites who had the esoteric knowledge and social right to use them in divination rituals. At the very least, it does appear that residents of earthen mounds produced mirrors for residents of masonry structures, as well as for their own domestic ritual.

Nevertheless, the status of mirror producers evidenced in artifact assemblages does appear to be relatively higher, but without many of the hallmarks of elite culture. None of the mirror producers in earthen mounds had masonry construction of their dwellings, tomb or cist-style burials, or certain jade artifacts such as carved plaques or earflares. Thus the material culture suggests different social identities for the producers and consumers of pyrite mirrors, possibly an emerging "middle class" of artisans. Some prestigious artifacts, such as figurines and imported ceramics suggest that there was some honor and social reward in the form of cultural capital associated with production of these artifacts. Elites derived power from the possession, use, and probably distribution of pyrite mirrors, which represented divinity, royalty, ritual potency, and creative power. Producers of these mirrors may have also been able to co-opt some of this power. They were likely able to use mirrors in their own domestic rituals, or at least identify with the creation of a powerful object, linking themselves to the divine act of creation.

REFERENCES

Ashmore, Wendy. 1980. "Discovering Early Classic Quirigua." *Expedition* 23: 35–44.

Barrientos, Tomás, Hector Escobedo, and Stephen D. Houston. 1997. "PN1: Excavaciones en la Estructura O-13." In *Proyecto Arqueológico Piedras Negras: Informe Preliminar No.1, Primera Temporada 1997*, edited by H. L. Escobedo and S. D. Houston, 1–20. Provo, UT: Brigham Young University.

Barrientos, Tomás, Brigitte Kovacevich, Michael Callaghan, and Lucía Morán. 2000. "Investigaciones en el area residencial sur y suroeste de Cancuén." In *Proyecto Cancuén: Informe Preliminar No. 2-Segunda Temporada*, ed. Arthur Demarest and Tomás Barrientos. Guatemala: Institute of Anthropology and History.

Bell, Ellen E. 2002. "Engendering a Dynasty: A Royal Woman in the Margarita Tomb, Copán." In *Ancient Maya Women*, ed. Traci Ardren, 89–104. Walnut Creek, CA: Altamira Press.

Besterman, Theodore. 1997. *Crystal Gazing: Study in the History, Distribution, Theory and Practice of Scrying*. Whitefish, MT: Kessinger Publishing.

Bill, Cassandra, and Michael Callaghan. 2001. "Frecuencias Relativas de los Tipos y Modos Cerámicos en Cancuén." In *Proyecto Cancuén: Informe Preliminar No. 3-Tercera Temporada*, ed. Arthur Demarest and Tomás Barrientos. Guatemala: Institute of Anthropology and History.

Bill, Cassandra, Michael Callaghan, and Jeannette Castellanos. 2003. "La Cerámica de Cancuén y la Región del Alto Pasión." In *Proyecto Cancuén: Informe Preliminar No. 4-Cuarta Temporada*, ed. Arthur Demarest and Tomás Barrientos. Guatemala: Institute of Anthropology and History.

Blainey, Marc G. 2007. "Surfaces and Beyond: The Political, Ideological, and Economic Significance of Ancient Maya Iron-Ore Mirrors." Master's thesis, Trent University, Peterborough, Ontario, Canada.

Braswell, Geoffrey, Christian M. Prager, and Cassandra R. Bill. 2005. ""The Kingdom of the Avocado: Recent Investigations at Pusilhá, a Classic Maya City of Southern Belize." Slovene Anthropological Society." *Anthropological Notebooks* 11: 59–86.

Bruhns, Karen O. 1989. "Crucible: Sociological and Technological Factors in the Delayed Diffusion of Metallurgy to Mesoamerica." In *New Frontiers in the Archaeology of the Pacific Coast of Southern Mesoamerica*, edited by Frederick Bove and Lynette Heller, 221–228. Anthropological Research Papers 39. Tempe: Arizona State University.

Callaghan, Michael, and Cassandra R. Bill. 2003. "Análisis Cerámico por Sector Residencial." In *Proyecto Cancuén: Informe Preliminar No. 5-Quinta Temporada*, ed. Arthur Demarest and Tomás Barrientos. Guatemala: Institute of Anthropology and History.

Callaghan, Michael, Cassandra Bill, Jeannette Castellanos, and Ronald Bishop. 2004. "Gris Fino Chablekal: Distribución y análisis socio-económico preliminar en Cancuén." In *XVII Simposio de Investigaciones Arqueologicas en Guatemala, 2003*, ed. Barbara Arroyo and Juan Pedro Laporte, 345–362. Guatemala: Museo Nacional de Arqueología y Etnología.

Carlson, Jonathan B. 1981. "Olmec Concave Iron-Ore Mirrors: The Aesthetics of a Lithic Technology and the Lord of the Mirror." In *The Olmec and Their Neighbors: Essays in Memory of Matthew W. Stirling*, ed. Elizabeth P. Benson, 117–147. New Orleans: Tulane University.

Carlson, Jonathan B. 1993. "The Jadeite Mirror: An Olmec Concave Jadeite Pendant." In *Precolumbian Jade: New Geological and Cultural Interpretations*, ed. Frederick W. Lange, 242–250. Salt Lake City: University of Utah Press.

Child, Mark, and Jessica Child. 1999. "PN 44: Excavaciones en el Baño de Vapor N-1." In *Proyecto Arqueológico Piedras Negras: informe preliminar 3*, ed. H. L. Escobedo and S. D. Houston, 191–196. Guatemala: Instituto Nacional de Antropología e Historia.

Ciudad Ruiz, Andres, Maria Josefa Iglesias, Jesús A. Pavón, Alfonso L. García-Gallo, and Jorge E. Chocon. 2003. "La Entidad Política de Machaquila, Poptun, en el Clásico Tardío y Terminal." In *Informe de la Segunda Temporada, 2003*, 236–348. Guatemala: Instituto de Antropología e Historia.

Coe, William R. 1959. *Piedras Negras Archaeology: Artifacts, Caches, and Burials*. University Museum of Pennsylvania Monograph 18. Philadelphia: University Museum, University of Pennsylvania.

Coe, William R. 1967. *Tikal: A Handbook of the Ancient Maya Ruins*. Philadelphia: University Museum, University of Pennsylvania.

Coggins, Clemency. 1975. "Painting and Drawing Styles at Tikal: An Historical and Iconographic Reconstruction." PhD diss., Harvard University, Cambridge, MA.

Coggins, Clemency. 1979. "A New Order and the Role of the Calendar: Some Characteristics of the Middle Preclassic at Tikal." In *Maya Archaeology and Ethnohistory*, ed. Norman Hammond and Gordon R. Willey, 38–50. Austin: University of Texas Press.

Coggins, Clemency. 1988. "The Manikin Scepter: Emblem of Lineage." *Estudios de Cultura Maya* 17: 123–58.

Cook, Duncan E., Brigitte Kovacevich, Timothy Beach, and Ronald Bishop. 2006. "Deciphering the Inorganic Chemical Record of Ancient Human Activity using ICP-MS: A Reconnaissance Study of Late Classic Soil Floors at Cancuén, Guatemala." *Journal of Archaeological Science* 33 (5): 628–640. http://dx.doi.org /10.1016/j.jas.2005.09.019.

Davis-Salazar, Karla L., and Ellen E. Bell. 1999. "Una comparación de los depósitos funerarios de dos mujeres de elite en la Acrópolis de Copán." In *XIII Simposio de Arqueología de Guatemala*, ed. Juan Pedro Laporte and Hector Escobedo, 1113–1128. Guatemala: Museo de Arqueología e Etnología.

Durán, Diego. (Original work published 1588) 1994. *The History of the Indies of New Spain, translated, annotated, and introduction by Doris Heyden.* Norman: University of Oklahoma Press.

Eberl, Markus, and Takeshi Inomata. 2001. "Maya Royal Headband (Sak Hunal) from Aguateca." *Mexicon* 23: 134–5.

Ekholm, Gordon F. 1945. "A Pyrite Mirror from Queretaro, Mexico." *Carnegie Institution of Washington, Notes on Middle American Archaeology and Ethnology* 2 (53): 178–181.

Ekholm, Gordon F. 1973. "Archaeological Significance of Mirrors in the New World." In *Proceedings of the International Congress of Americanists* (40 session, Rome and Genova, 1972) 1: 133–136.

Escobedo, Hector, and Carlos Alvarado. 1998. "PN 1: Excavaciones en la Estructura O–13." In *Proyecto Arqueológico Piedras Negras: Informe Preliminar No. 2 Segunda Temporada 1998*, ed. Hector Escobedo and Stephen D. Houston, 217–248. Guatemala: Instituto de Antropología e Historia.

Escobedo, Hector, and Marcelo Zamora. 1999. "PN 47: Excavaciones en la Estructura R–5." In *Proyecto Arqueológico Piedras Negras: Informe Preliminar No. 3 Tercera Temporada 1999*, ed. H. Escobedo and S. D. Houston. Guatemala: Instituto de Antropología e Historia.

Fialko, Vilma. 2000. "El Espejo del Entierro 49: Morfología y Texto Glífico." In *El Sitio Maya de Topoxte: Investigaciones en una isla del Lago Yaxha, Peten, Guatemala*, edited by W. W. Wurster, 144–149. Mainz am Rhein, Germany: P. von Zabern, KAVA.

Flannery, Kent. 1968. "The Olmec and the Valley of Oaxaca: A Model for Inter-regional Interaction in Formative Times." In *Dumbarton Oaks Conference on the Olmec*, edited by Elizabeth Benson, 129–135. Washington DC: Dumbarton Oaks.

Garber, James. 1993. "The Cultural Context of Jade Artifacts from the Maya Site of Cerros, Belize." In *Precolumbian Jade: New Geological and Cultural Interpretations*, ed. Frederick Lange, 166–172. Salt Lake City: University of Utah Press.

Gonlin, Nancy. 2007. "Ritual and Ideology among Classic Maya Rural Commoners at Copán, Honduras." In *Commoner Ritual and Ideology in Ancient Mesoamerica*, ed. Nancy Gonlin and Jonathan C. Lohse, 95–144. Boulder: University Press of Colorado.

Gonlin, Nancy, and Jon C. Lohse. 2007. *Commoner Ritual and Ideology in Ancient Mesoamerica.* Boulder: University Press of Colorado.

Gullberg, Jonas E. 1959. "Technical Notes on Concave Mirrors." In *Excavations at La Venta, Tabasco*, edited by P. Drucker, R. F. Heizer, and R. J. Squier, 280–283. Smithsonian Institution Bureau of American Ethnology, Bulletin 170. Washington, DC: United States Government Printing Office.

Harrison, Peter. 1999. *Lords of Tikal: Rulers of an Ancient Maya City*. London: Thames and Hudson.

Healy, Paul F., and Marc G. Blainey. 2011. "Ancient Maya Mosaic Mirrors: Function, Symbolism, and Meaning." *Ancient Mesoamerica* 22 (02): 229–244. http://dx.doi.org/10.1017/S0956536111000241.

Heizer, Robert, and Jonas Gullberg. 1981. "Concave Mirrors from the Site of La Venta, Tabasco: Their Occurrence, Minerology, Optical Description, and Function." In *The Olmec and Their Neighbors: Essays in Memory of Matthew W. Stirling*, ed. Elizabeth P. Benson, 109–116. Washington, DC: Dumbarton Oaks.

Hendon, Julia. 2000. "Having and Holding: Storage, Memory, Knowledge, and Social Relations." *American Anthropologist* 102 (1): 42–53. http://dx.doi.org/10.1525/aa.2000.102.1.42.

Houston, Stephen D., and Ernesto Arredondo. 1999. "PN 34: Excavaciones en el Patio 1 de la Acrópolis." In *Proyecto Arqueológico Piedras Negras: Informe Preliminar No.3, Tercera Temporada 1997*, ed. Hector L. Escobedo and Stephen D. Houston, 1–20. Guatemala: Instituto de Antropología e Historia.

Houston, Stephen, and Hector Escobedo. 1997. *Proyecto Arqueológico Piedras Negras: Informe Preliminar No. 1: Primera Temporada 1997*. Provo, UT: Brigham Young University.

Houston, Stephen, and Hector Escobedo. 1998. *Proyecto Arqueológico Piedras Negras: Informe Preliminar No. 2: Primera Temporada 1998*. Provo, UT: Brigham Young University.

Houston, Stephen, and Hector Escobedo. 1999. *Proyecto Arqueológico Piedras Negras: Informe Preliminar No. 3: Primera Temporada 1999*. Provo, UT: Brigham Young University.

Houston, Stephen, David Stuart, and Karl Taube. 2013. *The Memory of Bones: Body, Being, and Experience among the Classic Maya*. Austin: University of Texas Press.

Hruby, Zachary X. 2006. "The Organization of Chipped-Stone Economies at Piedras Negras, Guatemala." PhD diss., University of California, Riverside, CA.

Inomata, Takeshi, and Daniela Triadan. 2000. "Craft Production by Classic Maya Elites in Domestic Settings: Data from Rapidly Abandoned Structures at Aguateca, Guatemala." *Mayab* 11: 2–39.

Inomata, Takeshi, Daniela Triadan, Erick Ponciano, Estela Pinto, Richard E. Terry, and Markus Eberl. 2002. "Domestic and Political Lives of Classic Maya Elites: The Excavation of Rapidly Abandoned Structures at Aguateca, Guatemala." *Latin American Antiquity* 13 (3): 305–330. http://dx.doi.org/10.2307/972113.

Joyce, Thomas, J. Cooper Clark, and Eric S. Thompson. 1927. "Report on the British Museum Expedition to British Honduras." *Journal of the Royal Anthropological Institute* 57: 295–323.

Kesler, J., and J. Ascarrunz. 1982. *Lead-Zinc Mineralization in Carbonate Rocks. Central of Guatemala.* Documento Inerino del Ministerio de Energia y Minas del Gobierno de Guatemala.

Kidder, Alfred V. 1947. *The Artifacts of Uaxactun, Guatemala.* Publication 576. Washington, DC: Carnegie Institution of Washington.

Kidder, Alfred V., Jesse Jennings, and Edwin M. Shook. 1946. *Excavations at Kaminaljuyu, Guatemala.* Publication 561. Washington, DC: Carnegie Institution of Washington.

Kovacevich, Brigitte. 2002. "Producción y Distribución de Artefactos Líticos." In *Proyecto Cancuén: Informe Preliminar No. 4-Cuarta Temporada*, ed. Arthur A. Demarest and Tomas Barrientos, 95–106. Guatemala: Instituto de Antropología e Historia.

Kovacevich, Brigitte. 2006. "Reconstructing Classic Maya Economic Systems: Production and Exchange at Cancuén." PhD diss., Vanderbilt University, Nashville, TN.

Kovacevich, Brigitte. 2007. "Ritual, Crafting, and Agency at the Classic Maya Kingdom of Cancuén." In *Mesoamerican Ritual Economy: Archaeological and Ethnological Perspectives*, edited by E. Christian Wells and Karla Davis-Salazar, 67–114. Boulder: University Press of Colorado.

Kovacevich, Brigitte. 2011. "The Organization of Jade Production at Cancuén, Guatemala." In *The Technology of Maya Civilization: Political Economy and Beyond in Lithic Studies*, ed. Zachary X. Hruby, Oswaldo Chinchilla, and Geoffrey Braswell. Sheffield, UK: Equinox Publishing.

Kovacevich, Brigitte, Duncan Cook, and Timothy Beach. 2004. "Áreas de actividad doméstica en Cancuén: Perspectivas basadas en datos líticos y geoquímicos." In *XVII Simposio de Investigaciones Arqueológicas en Guatemala, 2003*, ed. Juan Pedro Laporte, Barbara Arroyo, Hector L. Escobedo, and Hector Mejia, 897–912. Guatemala: Museo Nacional de Arqueología y Etnología.

Kovacevich, Brigitte, Hector Neff, and Ronald L. Bishop. 2005. "Laser Ablation ICP-MS Chemical Characterization of Jade from a Jade Workshop in Cancuen, Guatemala. In *Laser Ablation ICP-MS in Archaeological Research*, ed. Robert J. Speakman and Hector Neff, 38-57. Albuquerque: University of New Mexico Press.

Kovacevich, Brigitte, Claudia Quintanilla, and Moises Arriaza. 2003. "Excavaciones en el Grupo N11." In *Proyecto Cancuén: Informe Preliminar No. 5-Quinta Temporada*, ed. Arthur A. Demarest, Tomas Barrientos, Brigitte Kovacevich, Michael G. Callaghan, and Luis F. Luin. Guatemala: Instituto de Antropología e Historia.

Kunen, Julie L. 2004. *Ancient Maya Life in the Far West Bajo: Social and Environmental Change in the Wetlands of Belize.* Anthropological Papers of the University of Arizona, No. 69. Tucson: University of Arizona Press.

Langley, James C. 1986. *Symbolic Notation of Teotihuacan: Elements of Writing in a Mesoamerican Culture of the Classic Period.* BAR International Series 313. Oxford: British Archaeological Reports.

Macri, Martha J., and Matthew G. Looper. 2003. *The Classic Period Inscriptions.* vol. 1. The New Catalogue of Maya Hieroglyphs. Norman: University of Oklahoma Press.

Mata Amado, Guillermo. 2003. "Espejo de pirita y pizarra de Amatitlán." In XVI *Simposio de Investigaciones Arqueológicas en Guatemala, 2002,* edited by Juan Pedro Laporte, Barbara Arroyo, Hector L. Escobedo, and Hector Mejia, 847–856. Guatemala: Museo Nacional de Arqueología y Etnología.

Matsumoto, Mallory. 2013. "Reflection as Transformation: Mirror-Image Structure on Maya Monumental Texts as a Visual Metaphor for Ritual Participation." *Estudios de cultura maya* 41: 93–128. Accessed on December 10, 2013. http://www .scielo.org.mx/scielo.php?script=sci_arttext&pid=S0185-25742013000100004&lng =es&tlng=en. http://dx.doi.org/10.1016/S0185-2574(13)71378-9.

Merwin, R., and George Vaillant. 1932. "The Ruins of Holmul, Guatemala." *Memoirs* 3(2). Cambridge, MA: Harvard University, Peabody Museum of American Archaeology and Ethnology.

Miller, Arthur G. 1973. *The Mural Painting of Teotihuacan.* Washington, DC: Dumbarton Oaks Research Library and Collection.

Miller, Mary Ellen, and Karl Taube. 1993. *An Illustrated Dictionary of the Gods and Symbols of Ancient Mexico and the Maya.* London: Thames and Hudson.

Mills, Barbara. 2008. "Remembering While Forgetting: Depositional Practices and Social Memory at Chaco." In *Memory Work: Archaeologies of Depositional Practice,* ed. Barbara J. Mills and William H. Walker, 81–108. Santa Fe: School of American Research Press.

MINORSA. 1977. *Estudio de Variabilidad, Minas de Oriente.* Documento Interno del Ministerio de Energia y Minas del Gobierno de Guatemala.

Morris, Earl, Jean Charlot, and Ann Morris. 1931. *The Temple of the Warriors at Chichén Itzá,* Yucatan. Publication 406. Washington, DC: Carnegie Institute of Washington.

Nordenskiöld, Erland. 1926. "Miroirs convexes et concaves en Amérique." *Société des américanistes de Paris* 18: 103–110.

Olson, Valerie F. 1984. "Olmec Magnetite Mirrors: Optical, Physical, and Chemical Characteristics." *Applied Optics* 23 (24): 4471–4476. http://dx.doi.org/10.1364/AO .23.004471.

Parsons, Lee Allen. 1986. *The Origins of Maya Art: Monumental Stone Sculpture of Kaminaljuyu, Guatemala, and the Southern Pacific Coast.* Studies in Pre-Columbian Art and Archaeology 28. Washington, DC: Dumbarton Oaks.

Pires-Ferreira, Jane W. 1976. "Shell and Iron-Ore Mirror Exchange in Formative Mesoamerica, with Comments on Other Commodities." In *The Early Mesoamerican Village*, ed. Kent V. Flannery, 311–325. New York: Academic Press.

Proskouriakoff, Tatiana. 1962. "The Artifacts of Mayapan." In *Mayapan, Yucatan, Mexico*, edited by H. Pollock, R. Roys, T. Proskouriakoff, and A. Smith. Publication 619. Washington, DC: Carnegie Institute of Washington.

Reents-Budet Dorie. 1994. *Painting the Maya Universe.* Durham: Duke University Press.

Reents-Budet Dorie. 1998. "Elite Maya Pottery and Artisans as Social Indicators." In *Craft and Social Identity*, edited by Cathy Lynne Costin and Rita Wright. Arlington, VA: Archaeological Papers of the American Anthropological Association No. 8. http://dx.doi.org/10.1525/ap3a.1998.8.1.71.

Rivera Dorado, Miguel. 1999. "Espejos mágicos en la cerámica maya." *Revista Espanola de Antropologia Americana* 29: 65–100.

Robertson, Merle Greene. 1974. "The Quadripartite Badge: A Badge of Rulership." In *Primera Mesa Redonda, Part 1 1973*, edited by M. G. Robertson, 77–93. Pebble Beach, CA: Robert Louis Stevenson School, Pre-Columbian Research.

Robichaux, Hubert Ray. 1995. "Ancient Maya Community Patterns in Northwestern Belize: Peripheral Zone Survey at La Milpa and Dos Hombres." PhD diss., University of Texas, Austin.

Robin, Cynthia. 2003. "New Directions in Classic Maya Household Archaeology." *Journal of Archaeological Research* 11 (4): 307–356. http://dx.doi.org/10.1023 /A:1026327105877.

Roys, Ralph L. 1933. *Book of Chilam Balam of Chumayel.* Publication no. 438. Washington, DC: Carnegie Institution of Washington.

Ruppert, Karl. 1935. *The Caracol at Chichén Itzá, Yucatan, Mexico.* Publication no. 454. Washington, DC: Carnegie Institution of Washington.

Sahagún, Bernardino de. 1959. *Florentine Codex: General History of the Things of New Spain, Book 9, The Merchants.* Ed. and trans. Arthur J. O. Anderson and Charles E. Dibble. Salt Lake City: University of Utah Press.

Sahagún, Bernardino de. 1963. *Florentine Codex: General History of the Things of New Spain, Book 11, Earthly Things.* Ed. and trans. Arthur J. O. Anderson and Charles E. Dibble. Salt Lake City: University of Utah Press.

Sahagún, Bernardino de. 1981. *Florentine Codex: General History of the Things of New Spain, Book 10, The People.* Ed. and trans. Arthur J. O. Anderson and Charles E. Dibble. Salt Lake City: University of Utah Press.

Sahagún, Bernardino de. 1985. *Historia general de las cosas de la Nueva España.* Mexico City: Editorial Porrúa.

Sanders, William T. 1990. *Secretaria de Cultura y Turismo.* vol. II. Excavaciones en el área urbana de Copán. Tegucigalpa, Honduras: Instituto Nacional de Antropología e Historia.

Schele, Linda. 1974. "Lords of Palenque;: The Glyphic Evidence." In *Primera Mesa Redonda de Palenque, Part 1, 1973,* edited by M. G. Robertson, 63–75. Palenque Round Table (1 session, 1973). Pebble Beach, CA: Robert Louis Stevenson School, Pre-Columbian Art Research.

Schele, Linda. 1976. "Accession Iconography of Chan-Bahlum in the Group of the Cross at Palenque." In *The Art, Iconography, and Dynastic History of Palenque,* edited by M. G. Robertson, 9–34. Palenque Round Table (2 session, 1974). Pebble Beach, CA: Robert Louis Stevenson School, Pre-Columbian Art Research.

Schele, Linda, and David Freidel. 1990. *A Forest of Kings.* New York, W.: Morrow and Co.

Schele, Linda, and Jeffrey H. Miller. 1983. *The Mirror, the Rabbit, and the Bundle: Accession Expressions from the Classic Maya Inscriptions.* Studies in Pre-Columbian Art and Archaeology No. 25. Washington, DC: Dumbarton Oaks Research Library and Collection.

Schele, Linda, and Mary Ellen Miller. 1986. *The Blood of Kings: Dynasty and Ritual in Maya Art.* New York: George Braziller.

Schellhas, Paul. 1904. "Representation of Deities of the Maya Manuscripts." *Papers of the Peabody Museum of American Archaeology and Ethnology* 4(1). Cambridge, MA: Harvard University.

Sears, Erin L. 2000. "Análisis Preliminar de las Figurillas de Cancuén." In *Proyecto Cancuén: Informe Preliminar No. 2-Segunda Temporada,* ed. Arthur Demarest and Tomás Barrientos. Guatemala: Institute of Anthropology and History.

Séjourné, Laurette. 1957. *Pensamiento y religión en el México antiguo.* Mexico: Fondo de Cultura Económica.

Sharer, Robert J., and Loa P. Traxler. 2006. *The Ancient Maya.* 6th ed. Stanford: Stanford University Press.

Shook, Edwin, and Alfred V. Kidder. 1952. "Mound E-III–3, Kaminaljuyu, Guatemala." Publication no. 596. Washington, DC: Carnegie Institution of Washington.

Smith, A. Ledyard, and Alfred V. Kidder. 1951. "Excavations at Nebaj, Guatemala." Publication no. 546. Washington, DC: Carnegie Institution of Washington.

Stone, Andrea Joyce, and Marc Zender. 2011. *Reading Maya Art: A Hieroglyphic Guide to Ancient Maya Painting and Sculpture*. New York: Thames and Hudson.

Stuart, David. 2010. "Shining Stones: Observations on the Ritual Meaning on Early Maya Stelae." In *The Place of Stone Monuments: Context, Use and Meaning in Mesoamerica's Preclassic Transition*, ed. Julia Guernsey, John E. Clark, and Barbara Arroyo, 283–298. Washington, DC: Dumbarton Oaks.

Sugiyama, Saburo. 1989. "Burials Dedicated to the Old Temple of Quetzalcóatl at Teotihuacán, Mexico." *American Antiquity* 54 (1): 85–106. http://dx.doi.org /10.2307/281333.

Sugiyama, Saburo, and Leonardo López Luján. 2007. "Dedicatory Burial/Offering Complexes at the Moon Pyramid, Teotihuacan." *Ancient Mesoamerica* 18 (1): 127–146. http://dx.doi.org/10.1017/S0956536107000065.

Taschek, Jennifer. 1994. *The Artifacts of Dzibilichaltun, Yucatan, Mexico: Shell, Polished Stone, Bone, Wood, and Ceramics*. Middle American Research Institute Publication no. 50. New Orleans: Tulane University.

Taube, Karl. 1992a. "Iconography of Mirrors at Teotihuacan." In *Art, Ideology, and the City of Teotihuacan*, ed. Janet C. Berlo, 169–204. Washington, DC: Dumbarton Oaks Research Library Publication.

Taube, Karl. 1992b. *The Major Gods of Ancient Yucatan*. Washington, DC: Dumbarton Oaks Research Library and Collection.

Taube, Karl. 2001. "Mirrors." In *Archaeology of Ancient Mexico and Central America: An Encyclopedia*, ed. Susan T. Evans and David L. Webster, 473–474. New York: Garland.

Taube, Karl, and Reiko Ishihara-Brito. 2013. "From Stone to Jewel: Jade in Ancient Maya Religion and Rulership." In *Maya Art at Dumbarton Oaks*, ed. Joanne Pillsbury, Miriam Doutriaux, Reiko Ishihara-Brito, and Alexandre Tokovinine, 134–153. Washington, DC: Dumbarton Oaks.

Thompson, J. Eric S. 1939. *Excavations at San Jose, British Honduras*. Publication no. 506. Washington, DC: Carnegie Institution of Washington.

Toscano, Salvador. 1970. *Arte Precolombino de Mexico y de la America Central*. 3rd ed. Mexico City: Universidad Nacional Autonoma de Mexico.

Tozzer, Alfred M. 1941. "Landa's Relación de las cosas de Yucatán." In Papers of the Peabody Museum of Archaeology and Ethnology No. 18. Cambridge, MA: Harvard University.

Triadan, Daniella. 2000. "Excavaciones en Estructura M8-4, Operacion 23, Suboperacion A." In *Informe del Proyecto Arqueologico Aguateca: La Temporada de Campo 1999*, ed. Eric Panciano, Takeshi Inomata, and Daniela Triadan, 54–62. Guatemala: Instituto de Antropología e Historia.

von Winning, Hasso. 1987. *La iconografía de Teotihuacan: Los dioses y los signos*. 2 vols. Mexico City: Universidad Autonoma de Mexico.

Wauchope, Robert. 1975. *Zacualpa, El Quiché, Guatemala: An Ancient Provincial Center of the Highland Maya*. Middle American Research Institute, Publication no. 39. New Orleans: Tulane University.

Webster, David, and Nancy Gonlin. 1988. "Household Remains of the Humblest Maya." *Journal of Field Archaeology* 15 (2): 169–190. http://dx.doi.org/10.1179/009346988791974484.

Wells, E. Christian. 2006. "Recent Trends in Theorizing Prehispanic Mesoamerican Economies." *Journal of Archaeological Research* 14 (4): 265–312. http://dx.doi.org/10.1007/s10814-006-9006-3.

Wells, E. Christian, and Karla Davis-Salazar, eds. 2007. *Mesoamerican Ritual Economies*. Boulder: University Press of Colorado.

Willey, Gordon R., Richard M. Leventhal, Arthur A. Demarest, and William L. Fash, Jr. 1994. *Ceramics and Artifacts from Excavations in The Copán Residential Zone*. Papers of the Peabody Museum of Archaeology and Ethnology. vol. 80. Cambridge, MA: Harvard University.

Woodbury, Richard, and Aubrey Trick. 1953. *The Ruins of Zacaleu, Guatemala*. 2 vols. Richmond: William Byrd Press.

Zamora, F. Marcelo. 2002. "La industria de la pirita en Aguateca durante el periodo Clásico Tardío." Licenciatura Thesis, Facultad de Ciencias Sociales, Departmento de Arqueología, Universidad del Valle, Guatemala.

5

Two metallic (iron ore) minerals widely used at Teotihuacan were pyrite and hematite. Large volumes of
both specular and earthy hematite were used for various purposes during the city's occupation. This chapter discusses the uses of hematite and pyrite in luxury
items that were mainly used by Teotihuacan elites in
sacred contexts and complex rituals. It also provides
an overview of their uses as abrasives and pigments in
other production processes.

Among the more common uses for specular hematite was creating a particular red surface finish characteristic of earthenware bowls and ceramic vases for ritual use starting in the Late Tlamimilolpan phase (AD
250–350). This use increased during the Xolalpan phase
(AD 350–550), at which point specular red painted vessels with thickly applied paint occur in a wide variety
of contexts, suggesting a high degree of availability for
the Teotihuacan people. In the subsequent Metepec
phase (AD 550–650), however, the paint layer with
hematite is thinner and less specular (Rattray 2001),
with the vessels themselves being limited to elite contexts. These changes in distribution are likely the result
of shrinking supplies of resources associated with the
general economic instability impacting Teotihuacan.

In addition to its use in pottery, the earthy variety
of hematite was used to obtain a distinctive red pigment common in Teotihuacan murals (Gómez et al. in
press). The oldest wall paintings that have been discovered so far date from the Tzacualli phase (AD 50–150)

*Identification and Use
of Pyrite and Hematite
at Teotihuacan*

Julie Gazzola, Sergio
Gómez Chávez, and
Thomas Calligaro

DOI: 10.5876/9781607324089.c005

107

and were composed on clay. Analyses have confirmed the use of this mineral as a pigment. The use of the specular variety in murals began during the Tlamimilolpan stage and continued through both the Xolalpan and Metepec phases (Magaloni 1996).

Pyrite and hematite were also important in the decoration of sculptures and masks, used primarily as incrustations in the eyes. Its use has been inferred in the manufacture of mirrors formed by tesserae in sheets adhering to slate supports. Pyrite was used in these distinctive ornaments, in the form of disks (identified as *tezcacuitlapillis*) carried on the lower back by the upper classes (possibly military and priests), as a kind of brooch or mirror (Cabrera and Cabrera 1991; Cabrera and Serrano 1999) and/or as a status distinction (see chapter 1, this volume).

Due to their physical and symbolic characteristics, it is probable that both pyrite and hematite were used in specific ritual activities. Slate disks found in the tunnel beneath the Pyramid of the Sun were decorated with incisions on one side and pyrite mirrors on the other side, supporting the inference of their use in sacred and ritual contexts. Recently, during the exploration process of the tunnel under the Temple of the Feathered Serpent, in the Citadel Complex (La Ciudadela), we were able to record and recover evidence of the use of both minerals in large quantities in a context that holds a meaning related to the underworld, further confirming the role both iron-ores played in important ritual.

It should be mentioned that due to the oxidization transformation processes of both of these iron ores, in many archaeological contexts neither mineral retains its metallic luster. They are found, in almost all cases, only as yellowish or reddish concretions or as powdery or earthy substances that remain poorly adhered to certain classes of special objects. Both pyrite and hematite suffer physical and chemical transformations that lead to other secondary minerals (Chaumeton 1987). This process has likely led many researchers to confuse them and mention them merely as pigments (Rubín de la Borbolla, 1947: 68, Cabrera and Serrano 1999: 371). The lack of specific analysis looking for such iron ores has often led to erroneous interpretations, because humidity and pH conditions typically break the ores down into other substances. We discuss these chemical transformations in more detail below.

PYRITE AND HEMATITE

Pyrite is a mineral of the sulfide group, consisting of sulfur and iron (FeS_2), whereas specular and earthy hematite is a ferric oxide (Fe_2O_3). The minerals have gold and silver metallic lusters, respectively, while the earthy hematite variety is reddish. Pyrite has a hardness of 6–6.5, while hematite is 5–6 on the Mohs scale. As reported by Panczner (1987), the closest deposits of pyrite and hematite to Teotihuacan are located in Guanajuato, Guerrero, Hidalgo, Jalisco, Mexico, Michoacan, Oaxaca, Puebla, Veracruz, and Queretaro. Although sourcing of these minerals has not been possible, we know that Teotihuacan established early commercial and political relations with many faraway sites for obtaining of green stones, quartz, slate, calcite, pigments, wood, salt, mica, marine resources, cotton, and cinnabar, among many other resources (cf. Gazzola 2004; Gómez and Gazzola 2007; Saint Charles et al. 2010; Rosales and Manzanilla 2011) (see chapters 1 and 3, this volume). We also discovered an earthy hematite deposit on Patlachique hill, perhaps the main source of this mineral from earlier times.

Hematite and pyrite typically transform into other, secondary minerals. Under oxidizing conditions, metallic or iron sulfides are converted into sulfates. In this way, the oxidation of pyrite results in the formation of a mineral known as jarosite, $KFe_3(OH)_6(SO_4)_2$, a potassium and yellow hydrated iron sulfate, for which hardness varies from 2.5 to 3.5. Jarosite has a rhombohedral crystalline structure, and granular, fibrous, powdery, scabs and scales, and is insoluble in water. In addition to the purely chemical transformations, sulfates may be reduced to sulfur or sulfides by biooxidation thanks to the action of certain bacteria, which, due to the anaerobic decomposition of organic matter by the bacteria, produce hydrogen that reacts with the sulfates. For the case of hematite, the original mineral is transformed by hydration processes into goethite α-FeO(OH), hydrated ferric oxide $Fe_2O_3H_2O$, or iron oxide and hydrogen (acquiring shades of yellow ocher, red ocher, and brown ocher, respectively), as well as limonite $FeO(OH)\cdot nH_2O$ in tones of brown, yellowish brown to red (Chaumeton 1987). The transformation of hematite and pyrite by a process of hydration to secondary minerals such as goethite, limonite, and jarosite, has led some researchers to confuse these yellow and red ocher minerals (usually on objects) for pigments, or some kind of adhesive, when in reality they are hematite or pyrite converted into secondary minerals.

THE ANCIENT USES OF PYRITE AND HEMATITE AT TEOTIHUACAN

Earthy hematite has been found in low concentrations in the form of a powder in many contexts at Teotihuacan. Only in a few areas, specifically in the Atetelco compound and in the Compound of the Glyphs (Conjunto de los Glifos) at the La Ventilla neighborhood, were the minerals found covering the entire floor surface. Unfortunately, it is not known what value or possible ritual significance hematite may have had in this kind of context. A significant amount of hematite apparently stored in the room of a house, excavated in the village of San Francisco Mazapa, might indicate that site as a place of preparation and distribution (Sánchez Morton 2011).

Earlier we mentioned that by using different analyses it has been possible to discover that the earthy variety of hematite was used as a pigment in the oldest mural painting (Gómez et al. in press) and we know that the specular variety was used extensively in later stages (Magaloni 1996). In addition, the earthy variety was used in the finished surface of ceramic vessels and figurines in the early stages (Gazzola, in preparation), and the specular variety from the Early Tlamimilolpan phase through the fall of Teotihuacan. Some evidence suggests that the hematite was used as an abrasive for lapidary production (Gazzola 2007), but there are also examples of objects (figure 5.1) that could be used as ornaments or decorations (Gómez 2000).

Due to its luster and reflective characteristics, pyrite was often used as incrustation in the eyes of masks (Gazzola 2009a) and greenstone sculptures (Cabrera and Sugiyama 1999: 31), as dental inlays, and on at least one figurine ornament in the burials of the Temple of the Feathered Serpent (Rubín de la Borbolla 1947: 69, Figures 10 and 14). Numerous slate disks located in the hip area of the skeletons of several individuals sacrificed for the consecration of the same temple may have been attached to laminar pyrite, as evidenced by the yellow substance found on them (Cabrera and Cabrera 1991: 24, Cabrera and Serrano 1999).

The earliest research concerning these minerals at Teotihuacan occurred three decades ago when Rodríguez (1982: 59) found two small pieces of metallic ore associated with a sculpture of Huehueteotl (old god of fire) in Structure 2E of the North Quad Complex of the La Ciudadela complex. The identification of these objects was possibly the first modern analysis made on this type of material, considered strange at the time of these earlier excavations. Since then, Sugiyama and López Lujan (2006: 30) have inferred its use in dedication offerings for the construction of the Pyramid of the Moon. During the explorations in this temple, several slate objects were found with traces of secondary minerals associated with the main offerings. A greenstone sculpture placed standing in the center of one of the slate disks covered with pyrite was part of this exceptional finding. Further analysis was possible through the discovery of a clay figurine, put on a slate disk with traces of jarosite along with an offering, and found during work to build the road exiting the Tulancingo pyramids.

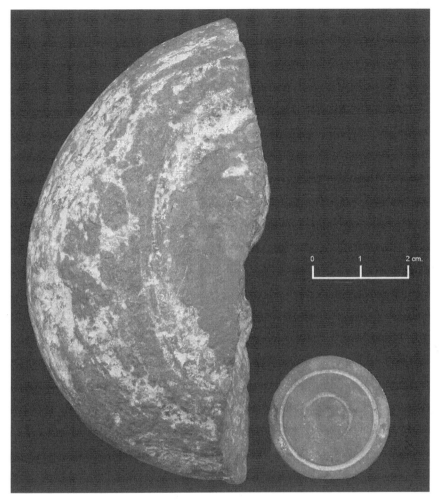

FIGURE 5.1. *Objects of metallic minerals such as pyrite ores have been worked in the lapidary workshops of La Ventilla (photo by M. Morales).*

Jarosite was also found associated with other clay figurines of females and children (Sánchez Rodríguez and Rubio 1997).

Recently, archaeologists recorded large slate disks associated with early substructures of the Pyramid of the Sun (Sugiyama, Sugiyama, and Sarabia 2013), which may also have been pyrite used as mirrors. In the tunnel discovered in 1973 under the same pyramid, some slate discs were found decorated with various designs on one side, while the other side was possibly used as a mirror

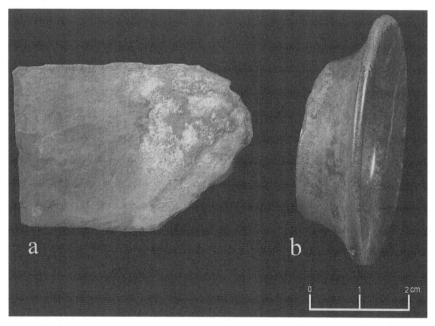

FIGURE 5.2. *(a) Basalt slab with yellow substance, probably jarosite, and (b) earplug of pyrite from the Structure 19 excavated by Gómez (photos M. Morales).*

formed by small plates of pyrite (Villa 2011).[1] Great worked slabs of the interior of the tunnel preserved remains of jarosite and iron oxide, suggesting that they could also have had plates of pyrite and hematite glued to them, probably to recreate the sky (Figure 5.2a).

Another of the few examples we can mention to illustrate the use of iron-ore minerals corresponds to a circular earplug manufactured in pyrite (Gómez 2002), found in a tomb in which imported materials suggest ties to western Mexico. This amazing piece could have been made somewhere in that region, which has a much older tradition in the use of pyrite and hematite, and the earplug therefore most likely was imported from there to Teotihuacan (figure 5.2b).

Recently while exploring the tunnel under the Temple of the Feathered Serpent (Gómez and Gazzola in press), we located a large amount of specular hematite scattered on the top floor of occupancy on the side walls and the roof duct, in addition to a large fragment of both products derived from pyrite and hematite minerals that were deposited as part of the fill closure. Many of the objects recovered in the form of disks and small tesserae made of slate have jarosite remains attached, so we assume that at one time the pieces had pyrite plates affixed to one side.

ANALYSIS FOR THE IDENTIFICATION OF MINERALS

In order to identify the two powdery substances (yellow and red) attached to various objects (initially assumed to have derived from solid iron-ores), analyses were performed by X-ray diffraction and by scanning electron microscope in the laboratories of the Mexican National Institute of Archaeology and History (INAH).

ANALYSIS BY XRD

The process of X-ray diffraction (XRD) used in the analysis of a powder simply requires a sample size of as little as a hundredth to a tenth of a milligram in order to obtain a diffractogram. The powder sample is placed on a sample holder, which is then mounted on a diffractometer with a counter. The interaction between the beam of photons and the crystallized material leads to a diffraction of the beam registered by the counter moving on a goniometer. The main objective of the analysis is to determine the crystal structure of the minerals present in the sample. The crystal lattice is constituted by families of parallel planes formed by atoms. These planes play the role of mirrors for the X-rays: "Each plane reflects only a small fraction of the beam, as if it were almost transparent. They are obtained from diffracted rays when reflections due to the parallel planes interfere additively" (Kittel 1972).[2] The spectrum of the XRD is a set of characteristic diffraction peaks, reflecting the minerals present in the analyzed sample, which can then be compared with data and records of the International Reference Library (JCPDS—International Center for Diffraction Data).

ANALYSIS BY SEM

The chemical characterization of specular hematite was carried out with the help of Ing. Gerardo Villa Sánchez, using a Jeol brand scanning electron microscope (SEM), with an energy dispersive spectrometer (EDS) for elemental chemical analysis. This is a nondestructive technique in which an electron beam is emitted on the material surface in order to analyze the various signals produced by interaction of the beam with the sample.

SEM analyses were performed on several samples of red mineral located on different objects. The major chemical elements identified and quantified

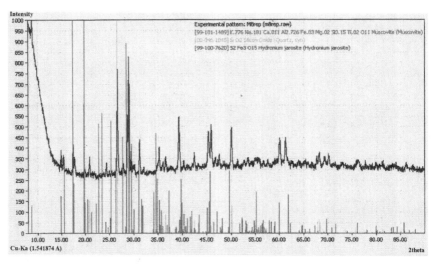

FIGURE 5.3. *XRD spectrum identifies the yellow substance as jarosite.*

by using an Oxford INCA processor were oxygen (48.37%) and iron (29.14%), capable of exhibiting characteristic features of the specular hematite (figure 5.3). The spectra obtained show a high presence of iron. However, since the material is degraded these results display a different mineral phase from pure pyrite and hematite. The yellow substance, analyzed by XRD, was identified as a combination of jarosite (figure 5.2), potassium, and iron sulfate, confirming the products of degraded pyrite; the red substance was identified as iron oxide, which corresponds precisely to the oxidation of the specular hematite identified by SEM (figure 5.3).

METALLIC MINERALS RECOVERED FROM THE TUNNEL UNDER THE TEMPLE OF THE FEATHERED SERPENT

It was noted earlier that during the exploration of the tunnel under the Temple of the Feathered Serpent, a large amount of specular hematite was found scattered on the floor, walls, and ceiling of the duct. We have also noted that many of the objects recovered have the remains of red and yellow substances identified as secondary minerals derived from the decomposition of pyrite and hematite. Among the slate objects (which originally had adhered pyrite and specular hematite) are over two hundred disks, some complete, along with many fragments. All have different diameters ranging from 2.5 to 30 cm and are between 2.5 to 4 mm thick. Full disks have a weight ranging from 5 to 44 grams.[3]

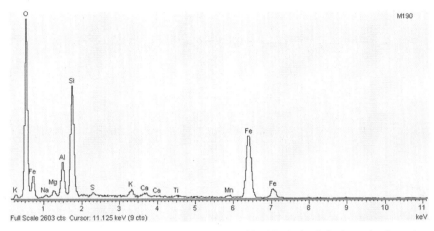

FIGURE 5.4. *Spectrum and table of results obtained by SEM of a disk of specular hematite.*

Most of the disks had several holes that were most likely used to tether the object with a cord, as a pendant. Some disks, identified by their dimensions as tezcacuitlapillis, present incised motifs (Gazzola 2009b) and cinnabar, while another example is completely impregnated with red ocher on one side and probably pyrite on the other. Other slate disks are smaller and unperforated. They could have been used as decorations or as pendants, rather than as tezcacuitlapillis, with pyrite or hematite attached to one side (see chapter 1, this volume).

We recovered thousands of small plates of various shapes and sizes. They ranged from square to rectangular, undulating, serpentine, circular, or trapezium in form and are beveled on their edges. All have several biconical perforations that begin on the back and end on the front. Usually one of the sides of these plates has a polished finish, while the other is rougher and possesses the yellow or red substances identified as iron oxide and jarosite (figure 5.6). It is likely that the plates formed part of necklaces, breastplates, or armor used by personages with a high social status.

Several kilograms of jarosite and iron oxides were recovered between the fillings positioned along the tunnel. Most have an irregular shape and are characterized by their fragility and powdery state, a situation associated with wet conditions present in the tunnel from its construction, which probably caused the transformation of the original minerals. Apparently all the fragments that we identified were deposited without any discernible order and the recovery of these objects has, at times, been complicated by their placement

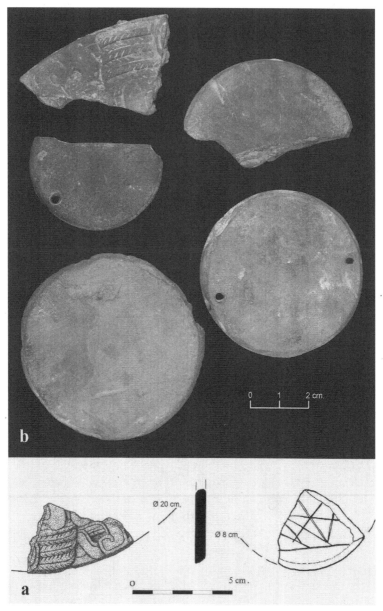

FIGURE 5.5. *Disks located in the tunnel, which illustrates the presence of Pre-Ciudadela compound 1 period in the tunnel. In all cases it is remarkable that the yellow substance is still adhering to the surface (drawing Fil de Coton photo by M. Morales).*

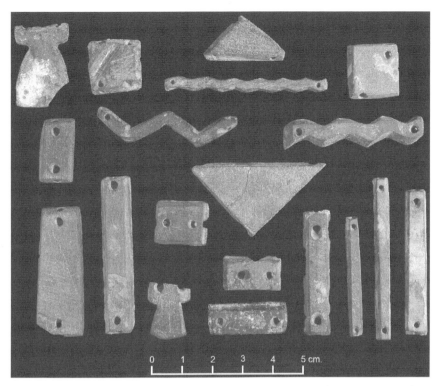

Figure 5.6. *Plates of slate with jarosite, from fillings of the tunnel (photo M. Morales).*

under the rocks and soil that make up the tunnel fill. It remains necessary to complete the exploration of the underground conduit to provide a comprehensive explanation of the presence of these materials, their meaning, and their ritual use.

MANUFACTURING OBJECTS WITH PYRITE AND SPECULAR HEMATITE

Due to the transformation process of pyrite and hematite under oxidation, only a few small articles made of either mineral have been preserved with their metallic luster intact. Nevertheless, and even without the presence of pyrite, we can deduce that several disks served as the basis of slate mirrors or *tezcacuitlapillis* and were used for ritual purposes or were carried by some individuals to display their social status. In fact, we note that most of the slate discs have been found in offerings of three major architectural monuments at Teotihuacan: the pyramids of the Sun and the Moon and the Temple of the Feathered Serpent. Such prominent proveniences undoubtedly indicate the symbolic import and sacredness of these objects.

In the case of so-called clasp mirrors, or *tezcacuitlapillis*, it is difficult to know what techniques and tools were used in the working of these pyrite and hematite specimens because of the bad conservation of the materials. However, it is likely that the working of both materials was done with techniques and tools similar to those for lapidary production (see chapters 3 and 4, this volume). The study of two conserved objects discovered in the lapidary workshops of La Ventilla gives us evidence about the techniques used in the working of both minerals. The technique used seems to have been a wearing process of gradual and continuous effort to reach the desired shape (Lorenzo 1965). Evidence has shown that this technique can be used for cutting, perforating, abrading, polishing, and buffing (Gómez and Gazzola 2011).

One of the La Ventilla objects is a mineral block with traces of cutting by abrasion. The second is a small disk of 3 mm thickness and approximately 2.5 cm in diameter with two concentric incised lines (figure 5.1). In the first piece, cuts were apparently performed using a fiber rope, some kind of abrasive material, and a water carrier. In the small disk, a tubular instrument was also employed in a rotating motion on the circular plate using a fine abrasive to achieve two concentric, nearly perfect circles. The topcoat must have served both as a polishing and burnishing agent to achieve a homogeneous surface, which in the case of pyrite may have been the most effective method.

In these workshops a small plate of basalt used as a working platform was found covered with hematite powder. This item is located directly on the floor of a room surrounded by other tools and a concentration of the mineral. Due to the natural hardness of hematite, we also proposed its use as an abrasive for cutting other materials (Gazzola 2007; see also chapters 2 and 3, this volume).

Objects with pyrite and specular hematite were likely made by full-time craftspeople in specialized workshops, some of which were probably under state control (Gazzola 2010). We recognized four stages in the process of the slate disks' manufacture. The first stage consisted in the separation of the slate sheets according to the required thickness. Next, the sheets were separated by indirect percussion with the help of a chisel of bone (Romero Hernández 2004) or wood. Once the plate was obtained, a "pattern" was used to guide the cutting of the sheet into a circle with a sharp tool of stone or hard bone. Finally, the edges were beveled by abrasion techniques to give them a smooth surface (figure 5.7c).

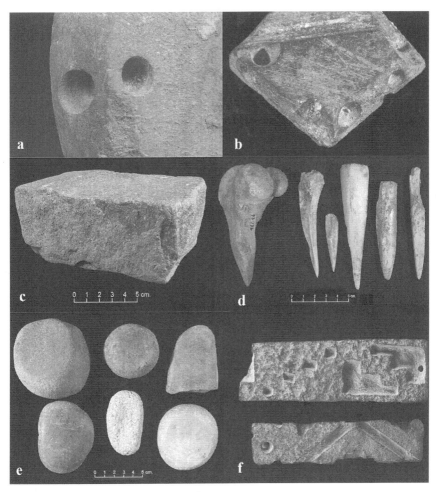

FIGURE 5.7. *(a) Side–by–side conical perforations in a slate disk, (b) perpendicular conical perforations in a slate plate, (c) table of abrasion, (d) bone tools for drilling, (e) stone tools for polishing and burnishing, (f) two views of an incised plate to receive the decoration (photos by M. Morales).*

Most of the slate that was the basis for the mirrors and ornaments has biconical (figure 5.7A) or conical perforation, either side-by-side or perpendicular (figure 5.7b). Indirect percussion initiated the work by using a bone drill (figure 5.7d) to make a small depression to mark the future location of the perforation and to prevent slippage of the perforating tool. The abrasive material used to start the hole had to be fine sand, and with the gradual enlargement of the diameter the maker probably used coarser abrasives. In biconical perforations, it is likely that as drilling progressed toward the center of the object, drilling was started on the opposite surface. A thin bone, such as a needle, and a fine abrasive were employed to join together the two points on each side of the drilling.

Abrasives could be thick or thin according to the manufacturing stage and among these compounds stone workers may have used crushed or powdered obsidian, quartz sand, limestone fragments, or hematite. With the small plates localized in the tunnel under the Temple of the Feathered Serpent, the position of the holes on many of these objects indicates that they were probably tied or sewn to a fabric, so that pyrite or hematite must have been affixed with an adhesive preparation such as copal or orchid sap, for example.

A third stage consisted of polishing (figure 5.7e), by which the manufacturing tracks are erased. This was achieved by polishing with a smooth stone or by rubbing the object on a table with abrasives. The fourth stage of production involved the addition of decoration on one or both sides of the object, which in some cases included incision in low relief, inserting pigment such as cinnabar or hematite, or gluing sheets of pyrite and hematite to the surface (figure 5.7f) with an adhesive of organic origin.

CONCLUSIONS

Pyrite and hematite are two metallic minerals that had a much more extensive and varied use at Teotihuacan than at first might be supposed. From the early stages, earthy hematite was used as a pigment in the creation of murals, while the specular variety and pyrite were used mainly in ritual contexts. From the Tlamimilolpan phase, specular hematite had more frequent use in the achievement of a distinctive surface finish of painted pottery and mural painting. Because of its hardness, hematite could also be used as an abrasive in lapidary and stone crafting.

In Teotihuacan mural paintings (Miller 1973), there are representations of circular elements ornamented with feathers that are depicted as being secured to the hip by individuals richly dressed. These elements are probably pins, or *tezcacuitlapillis*. The symbolism of the reflective material of both minerals implies a deeper meaning and practical use as omens and in divination, for example. Without a doubt, we need to further understand the meaning of these objects at Teotihuacan. For now we can assume, such as in the case of the ancient Maya (Healy and Blainey 2011), that pyrite mirrors allowed access to otherworldly spiritual dimensions (see chapters 9 and 10, this volume).

The transformation processes by hydration and oxidation of both pyrite and specular hematite lead to the formation of secondary minerals. Jarosite and limonite are the result of chemical and physical transformations of pyrite, whereas goethite results from the degradation of hematite. The analyses conducted on samples of both minerals corroborated that the yellow and red substances impregnated on the surface of many objects were originally adhered pyrite and hematite plates or inlays. Although the chemical transformation may result from oxidation and hydration, the possibility exists that in the case of pyrite degradation, this change would have resulted in jarosite through the process of biooxidation, thanks to the action of bacteria present in the adhesive compound of an organic material that has yet to be identified.

The analyses discussed above have confirmed the hypothesis that yellow and red substances adhering to sculptures (mainly eyes) and ornaments (such as *tecazcuitlapillis*) are not pigments or adhesives as some researchers have mentioned, but may actually be decomposed iron-ore minerals originally used for their reflective qualities. Over time and through chemical processes (such as oxidation and hydration), these reflective stones have been transformed into jarosite, goethite, and limonite, the form in which we find them in archaeological contexts.

Acknowledgment: Thanks to Erika Begun and Claudia Garcia des Lauriers for providing the English corrections to this chapter.

NOTES

1. According to T. Villa, limonite was identified.

2. "Le but de l'analyse est de déterminer la structure crystalline des minéraux présents dans l'échantillon. Le réseau du crystal est constitué par des familles de plans parallèles que forment les atomes. Ces plans jouent le rôle de miroirs vis-à-vis des rayons X incidents. Chaque plan réfléchissant seulement une petite fraction du rayonnement, comme s'il est quasi transparent. On obtient de rayons diffractés que lorsque les réflexions dues aux plans parallèles interfèrent de façon additive."

3. Larger disks are heavier, but as they are fragmented, we did not consider them.

REFERENCES

Cabrera, Rubén, and Oralia Cabrera. 1991. "El proyecto Templo de Quetzalcóatl." *Arqueología* 6, 2nd ed.: 19–31.

Cabrera, Rubén, and Carlos Serrano. 1999. "Los entierros de la Pirámide del Sol y del Templo de Quetzalcóatl, Teotihuacán." In *Prácticas funerarias en la Ciudad de los Dioses*, edited by Linda Manzanilla and Carlos Serrano, 345–397. México: UNAM, IIA.

Cabrera, Rubén, and Saburo Sugiyama. 1999. "El Proyecto Arqueológico de la Pirámide de la Luna." *Arqueología* 21: 19–33.

Chaumeton, Hervé. 1987. "Les Minéraux, petit guide encyclopédique." Paris: Artémis edition.

Gazzola, Julie. 2004. "Uso y significado del cinabrio en Teotihuacán." In *Memoria de la Segunda Mesa Redonda de Teotihuacán, La costa del Golfo en tiempos teotihuacanos propuestas y perspectivas, Premio*, ed. G. María Elena Ruiz and S. Arturo Pascual, 541–569. México: INAH.

Gazzola, Julie. 2007. "La producción de cuentas en piedras verdes en los talleres lapidarios de La Ventilla, Teotihuacán." *Arqueología* 36: 52–70.

Gazzola, Julie. 2009a. "Arquitectura de fases tempranas en Teotihuacán." *Arqueología* 42: 216–233.

Gazzola, Julie. 2009b. "La cadena operativa en la fabricación de mascaras en los talleres de lapidaria de La Ventilla, Teotihuacán." In *Reflexiones sobre la industria lítica*, coordinated by Leticia González A. and Lorena Mirambell, 61–77. Colección Científica 561. Mexico: INAH.

Gazzola, Julie. 2010. "Producción en Teotihuacán: taller de barrio y taller estatal." In *Técnicas y Tecnología en el México Prehispánico*, ed. Rosalba Nieto Calleja, 35–51. Mexico: INAH.

Gazzola, Julie. (In preparation). "La cerámica de las fases Tzacualli y Miccaotli en Teotihuacán: El caso del Conjunto 1 Preciudadela." *INAH*.

Gómez, Sergio. 2000. *La Ventilla: Un barrio de la antigua ciudad de Teotihuacán. Exploraciones y resultados*. Licenciatura thesis, Escuela Nacional de Antropología e Historia, Mexico.

Gómez, Sergio. 2002. "Presencia del occidente de México en Teotihuacán. Aproximaciones a la política exterior del Estado teotihuacano." In *Memoria de la Primera Mesa Redonda de Teotihuacán, Ideología y política a través de materiales, imágenes y símbolos*, ed. Elena Ruiz Gallut María, 563–625. México: INAH-UNAM.

Gómez, Sergio, and Julie Gazzola. 2007. "Análisis de las relaciones entre Teotihuacán y el occidente de México." In *Dinámicas culturales entre el occidente, el centro-norte y la cuenca de México, del Preclásico al Epiclásico*, coordinated by Brigitte Faugère, 113–135. Mexico: El Colegio de Michoacán, CEMCA.

Gómez, Sergio, and Julie Gazzola. 2011. "La producción lapidaria y malacológica en la mítica Tollán-Teotihuacán." In *La producción artesanal y especializada en Mesoamérica. Áreas de actividad y procesos productivos*, edited by Linda Manzanilla and Kenneth Hirth, 87–130. México: INAH, UNAM, IIA.

Gómez, Sergio, and Julie Gazzola. In press. "Cosmogonía y grupos de poder. Escenificaciones rituales y políticas en el santuario de La Ciudadela, Teotihuacán." In *BAR*. Paris.

Gómez, Sergio, Julie Gazzola, and Javier Vázquez. In press. "La conservación de pintura mural sobre barro en Teotihuacan." In *Memorias de la V Mesa Redonda de Teotihuacán: investigaciones recientes, centro y periferia*. México City: INAH.

Healy, Paul, and Marc Blainey. 2011. "Ancient Mosaic Maya Mirror: Function, Symbolism and Meaning." *Ancient Mesoamerica* 22 (2): 229–244. http://dx.doi.org /10.1017/S0956536111000241.

Kittel, C. 1972. *Introduction à la physique de l'état solide*, 3rd ed. Paris: Dunod.

Lorenzo, José Luis. 1965. *Los artefactos de Tlatilco*. Serie Investigaciones 7. México: INAH.

Magaloni, Diana. 1996. "Materiales y técnicas de la pintura mural maya." In *La pintura mural prehispánica en México I, Teotihuacán, t. II, Estudios,* coordinated by Leticia Staines C, 187–225. México: IIE-UNAM.

Miller, G. Arthur. 1973. *The Mural Painting of Teotihuacan.* Washington, DC: Dumbarton Oaks, Trustees for Harvard University.

Panczner, W. D. 1987. *Minerals of Mexico*. New York: Van Nostrand Reinhold. http:// dx.doi.org/10.1007/978-1-4757-5848-1.

Rattray, Evelyn. 2001. *Cerámica, cronología y tendencias culturales*. Serie Arqueología de México. México: INAH, Universidad de Pittsburgh.

Rodríguez, G. Ignacio. 1982. "Frente 2." In *Memoria del Proyecto Arqueológico Teotihuacán 80–82,* coordinated by Rubén Cabrera C., Ignacio Rodríguez G., and Noel Morelos G., vol. 1: 55–73. Colección Científica 132. México: INAH.

Romero Hernández, Javier. 2004. *La industria ósea en un barrio teotihuacano. Los artefactos de hueso de La Ventilla*. Licenciatura thesis, Escuela Nacional de Antropología e Historia, INAH, México.

Rosales Edgar, and Linda Manzanilla. 2011. "Producción, consumo y distribución de la mica en Teotihuacán. Presencia de un recurso alóctono en los contextos arqueológicos de dos conjuntos arquitectónicos: Xalla y Teopancazco." In *La producción artesanal y especializada en Mesoamérica. Áreas de actividad y procesos productivos,* ed. Linda Manzanilla and Kenneth Hirth, 131–152. México: INAH, UNAM, IIA.

Rubín de la Borbolla, Daniel F. 1947. "Teotihuacán: Ofrendas de los templos de Quetzalcóatl." In *Anales del Instituto Nacional de Antropología e Historia,* vol. 2, 1941–1946, 61–72. México: INAH.

Saint-Charles Z., Juan Carlos, Carlos Viramontes A., and Fiorella Fenoglio L. 2010. *El Rosario, Querétaro: Un enclave teotihuacano en el Centro Norte. Tiempo y Región.* In *Estudios Históricos y Sociales* vol. 4. Mexico: Municipio de Querétaro, INAH, Universidad Autónoma de Querétaro.

Sánchez Morton, Ligia. 2011. "Acopio y suministro de pigmentos a nivel de barrio." Paper presented at the V Mesa Redonda de Teotihuacán, *Investigaciones recientes: Centro y periferia*. México: INAH.

Sánchez Rodríguez, Ernesto A., and Jaime Delgado Rubio. 1997. "Una ofrenda cerámica al este de la antigua ciudad de Teotihuacán." *Arqueología 18*, Segunda Época: 17–23.

Sugiyama, Saburo, and Leonardo López Lujan. 2006. "Sacrificio de consagración en la Pirámide de la Luna, Teotihuacan." In *Sacrificio de consagración en la Pirámide de la Luna, Teotihuacan*, edited by Saburo Sugiyama and Leonardo López Lujan, 25–52. Museo de Templo Mayor. México: Arizona State University, INAH.

Sugiyama, Nawa, Saburo Sugiyama, and Alejandro Sarabia. 2013. "Inside the Sun Pyramid at Teotihuacan, Mexico: 2008–2011 Excavations and Preliminary Results." *Latin American Antiquity* 24(4): 403–432.

Villa, Tomás. 2011. "Los Tezcacuitlapilli de la Pirámide del sol, condiciones sociales de producción y relaciones sociales asimétricas." Paper presented at the V Mesa Redonda de Teotihuacán, *Teotihuacán: investigaciones recientes. Centro y periferia*, México.

6

On How Mirrors Would Have Been Employed in the Ancient Americas

José J. Lunazzi

I made efforts to envisage how low-reflectivity (20%) polished-stone artifacts (i.e., "mirrors") made in Central and South America would have been employed in ancient times, and performed some experiments to prove the suggested possibilities. For example, I describe how it would be possible to make images of objects with an intense light source (mainly by direct sunlight), with the mirror located in its shadow, and how curvature affects the luminous intensity and the capability of making images of large objects. I also consider the possible use of mirrors for communicating at long distances through reflections. Some examples are given to consider the sharpness of the image as of good quality. The mirrors' ability to concentrate the sun to make fire is also discussed, making some experimental simulations. Some possibilities of mountings to hide the object to give a phantom impression are also considered.

As the chapters in this volume indicate, reflective optical elements are a proof of a sophisticated culture in action. Such objects are mentioned as being employed in several ancient cultures (Andersons 2007; Enoch 2006). The Egyptians used reflective optical elements to illuminate the inside of tombs, or to give the impression of mystery and supernatural power (Enoch 1999). In ancient Greece, Archimedes suggested burning enemy ships by concentrating the reflection of mirrors (Simms 1977). Mirrors are also mentioned as a product of ancient China (Xiu 1996).

DOI: 10.5876/9781607324089.c006

None of these functions for mirrors seem to have been present among the American cultures, well known by the middle of the twentieth century after findings in Mexico and Peru (Muelle 1940; Nordenskiöld 1926). Of these possibilities the refractive ones seems to be the least probable, because spherical polished quartz objects are rare. Mirrors, on the other hand, are part of Olmec iconography (Blainey 2007), are present in many archaeological contexts, and are shown in public collections. Although numerous publications have given consideration to archaeological mirrors, very few show the images these artifacts are capable of producing (Lunazzi 2007, 1995, 1996; Saunders 1988).

Unfortunately, the cultural remains give little information, so the tradition of making or employing these objects has been lost over time. It would be worth trying to rediscover the path of creation by allowing direct descendants of archaeological populations to play with similar elements, with a cultural heritage and close proximity to the living conditions of those ancient civilizations (see chapters 2 and 3, this volume).

My imagination was spurred by observing children playing with mirrors or lenses, and I recall scenes from Western movies in which Native Americans employed mirrors to make light signals with the sun. The author is reminded of a report from more than half a century ago that an amusement park in Buenos Aires, Argentina, had an attraction where mirrors were used to make a flower appear as a phantom image over a real flowerpot named "La Rosa Azteca" (Rebollo 1988). Seeing the sunlight reflected on the water when flying in a commercial airplane was also intriguing because of the physical similarity between water and low-reflectivity mirrors. One must also consider that ancient people could have conceived of a variety of techniques for optimizing the use of reflective objects. Indeed, in complex civilizations like those of ancient Mesoamerica, lithic specialists would have had enough time and interest to explore every possibility.

CONSIDERATIONS INVOLVING THE PHYSICS OF ARCHAEOLOGICAL MIRRORS

To locate the subject technically, we must consider that reflectivity is the measure of the amount of light energy that does not enter a material but that is bounced back toward the medium from which the light was coming, making symmetrical angles with the surface. If the light that is bounced back is spread through a large angle, then it is said to be dispersion reflected. A mirror of good quality has a low amount of dispersive reflection; the light that is not reflected in stone mirrors is mostly absorbed as heat. Reflectivity increases when the incidence tends to be grazing (at a few angular degrees from the surface), but it is not necessary to consider this in imaging because the common use of mirrors happens with an incidence of light rays between 45 and 90 degrees to the surface. What gives the mirror its capability to reflect is the presence of electrical charge on matter, even if its total sum is neutral. Reflectivity rises above 4% when the mobility of the electrons exists (what is called conductivity), with the material being an electricity conductor, as is the case with metals. The presence of metal in a mineral makes it proper for mirror making. If the surface is plane it just redirects the incident light giving symmetrical images, if convex it increases the divergence of the light to a larger angle and makes reduced images, which are always behind the mirror and upwards. A concave surface converges light to a focus from which it is diverged afterwards, but if an object is close to it, its image is upward and behind the mirror, as in the convex case, but enlarged. When placed far from the reflective surface the image becomes inverted and floats in front of the mirror, being reduced in size compared to that of the original object.

The archaeological mirrors I have seen and had the opportunity to examine personally had low reflectivity of between 5% (corresponding to anthracite [Calvo and Enoch 2007; Larsen 1979]), and 20% (corresponding to hematite [Carlson 1981]). Obsidian and other materials have a similarly low reflectivity. While pyrite mirrors could have as much as 60% reflectivity (figure 6.1), I could not see those because they are not shown at public museums in the capitals of Mexico and Peru. Even those that are shown to the public are not mounted in a position that makes it possible to reflect images in their surfaces, which requires much illumination. Considering how marvelous these objects are and the meaning of power they presumably transmitted to the ancient audience, how these technologies would have been employed by their owners is a matter of consideration. For instance, after the initial grinding, the surfaces of Olmec jade carvings were further finished by sanding and polishing, with the finest abrasive being used for the final, mirror-like polish (Taube 2004: 24). We know that the iron-ore hematite is currently used as polishing rouge, making an intriguing technical similarity between the very ancient techniques and present-day ones. The analysis of workmanship on other materials, such as magnetite, was also reported (Carlson 1993). The measurement of iron-ore's optical properties and its material characteristics was carried out in a few cases, such that important parameters like reflectivity can be obtained from general tables made for the materials in geology (Calvo and Enoch 2007; Olson 1984).

FIGURE 6.1. *Image of a hand in a plane mirror from Cupisnique culture, Museum Larco, Peru (photograph by José J. Lunazzi).*

Regarding their size, mirrors can be as small as a thumbnail (Heizer 1967). For experiments performed by my students and myself, a size of 15 cm was considered a maximum, even though this can be only an average size for ancient Maya mirror artifacts (Healy and Blainey 2011). Regarding their curvature, they can be plane, concave, or convex, although convex mirrors are rare and I know of only one example (Lunazzi 1996). According to the classification by Elkholm (Olivier 2004), we can identify three categories of Mesoamerican mirror: (1) the concave specimens of the Olmecs (within which the convex ones should be included, in my opinion) (figure 6.2); (2) pyrite and marcasite mirrors of the Classic and Postclassic periods; and (3) obsidian mirrors of the Postclassic period. In the future, more detailed analyses could be made, among which it would be useful to include ancient specimens from South America (present in museums at Peru), which are all plane mirrors (Lunazzi 2007).

FIGURE 6.2. *Reflected image of a human face in a convex Olmec mirror (photograph by José J. Lunazzi).*

ON THE MIRRORS' APPLICATIONS

Many applications have already been reported for the mirrors. As a social distinction, it was said that during the Locona phase in Chiapas (1400–1250 BC) (Taube 2004: 6), there is evidence of a chiefdom level of social stratification in which—unlike big man societies—high social status was inherited rather than achieved. A Locona-phase burial from El Vivero contained the remains of a child wearing a circular mica mirror on its forehead, quite probably a sign of high rank. It has also been noted that the San Lorenzo sculpture known as Tenochtitlan Monument 1 portrays a ballplayer employing a mirror pectoral atop a bound captive (Taube 2004, citing Bradley and Joralemon 1993). Mirrors were also worn as pectoral adornments by Olmec ballplayers (Taube 2004:11, 13, 14). The following section examines a list of the possible uses of low-reflectivity stone mirrors by ancient Mesoamericans:

1. To observe one's own face.
2. To observe flipped images, symmetrical in horizontal or vertical directions, and in depth.
3. To observe converging images that appear in front of concave mirrors.

4. To impress people with a bright reflection of the sun as a symbol of the divine.
5. To make fire.
6. To communicate by means of sunlight.

1. TO OBSERVE ONE'S OWN FACE

It is natural to conceive that the person who first polished a dark piece of stone had soon realized how the sun makes it shine, and how his/her face illuminated under the sun can be seen as reflected. The more it was polished, the better the effect, adjusting for the best possible results. The dark background of the material compensates for the low reflectivity, keeping the necessary contrast of the images. If polishing was accomplished over a horizontal support in an open area, as it is more reasonable to suppose, at some hour of the day when the sun rises or at dawn, if the worker was on the opposite side of the mirror, its rays would have been reflected. In that same position, exploring the mirror's angles leads to a position where the reflection is plainly illuminated by the sun while the mirror surface receives no light from it (this is the ideal, to have the better contrast for an effective image). The ancient person could not see his/her reflection under less-intense illumination (Lunazzi 1995, 1996, 2009a).

Without the mirrors, the only way that the ancients had for observing their reflections was in a deposit of calm water. Because the face must be horizontally placed, it makes it difficult for one to be plainly illuminated under the sunlight, even at dawn. The presence of wind destroys the image by agitation of the water's surface. To have the best conditions, the bottom must be dark. An artifact could have been made by putting water in a large black painted vessel. Even so, the reflectivity coefficient of light would be less than 4%, making the stone mirrors highly superior. With the stone mirrors on the other hand, a person can see his/her reflection even at midday or under less-intense illumination. Would the possibility of observing one's reflection have inspired the creative development of maquillage? Would the convex mirrors, being smaller and lighter and requiring less material and workmanship to produce, have been more popular? Because the details on a reflection are less clearly seen, my guess is that this could have happened only if observing one's own face had some kind of magical or spiritual significance. On the other hand, would the concave mirrors, showing a reflection in better detail through enlargement, be preferred? It must still be proved whether the image quality of these mirrors can be compared to that of mirrors actually being employed by women nowadays, and if so, this could be one of the reasons for the conversion of convex mirrors to concave ones, as indicated by Carlson (1981). Convex mirrors are a subproduct of mirror fabrication, and although it is easier to make the surface concave when starting from a plane stone, a lack of stone material would be a reason for the maker to prefer the convex pieces.

One must now give attention to one of the most impressive properties of ancient plane mirrors: because of their large size, the vision can be binocular and the image appears behind the surface of the mirror, in a place where the mirror itself, although thick, does not exists. Ancients may well have conceived of such optical phenomena as the revealing of a fantastic parallel universe co-existing with the human realm, as if the mirror connected two otherwise separate domains. Even modern literature (Borges 1934) and movies (Cocteau 1950) include this idea.

2. TO OBSERVE FLIPPED IMAGES, SYMMETRICAL IN HORIZONTAL OR VERTICAL DIRECTIONS, AND IN DEPTH

Perhaps it is not easy to notice at first that if, for example, a person looking into a mirror closes one eye, the eye of his/her mirror image appears to close the eye on the opposite left-right side. But if we come with one hand from one side to the plane mirror, we see a symmetrical hand coming to it both reaching one another at the mirror. This property was certainly observed by ancient mirror owners but it would be hard to identify this in iconography representing the mirror. The idea of this symmetry can arrive through many natural representations when an artist makes drawings. For instance, on a wall at the Mexican archaeological site of Teotihuacan there is the drawing of a jaguar symmetrically connected to its flipped image, side to side. Are there more examples of these horizontally flipping figures and would they be related to the properties of the mirrors? Furthermore, by putting one hand vertically on the opposite side of a horizontal mirror we can also notice the image being upside down, inverted in the vertical direction. The case of inverted upside down or vertically flipped images is not easy to find in the ancient iconography and also difficult to identify as related with mirrors. However, as we see in figure 6.3, some faces were drawn upside down (Carlson 1981, Figure 35), one image shows the middle lower part where five heads appear inverted; two are larger and located on the sides, while three are centered. Two other inverted heads were drawn in a similar position as can be seen in Carlson (1981, Figure 36). The curious fact is that they are smiling, a situation that is quite different

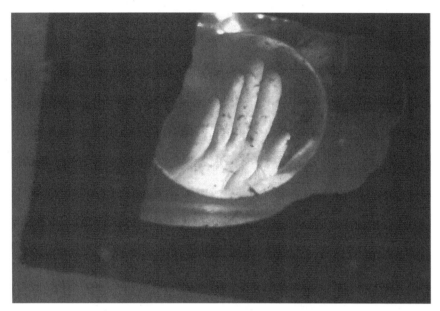

FIGURE 6.3. *The image of a hand can be seen as a convergent image appearing in front of an Olmec mirror if two-color 3D glasses are employed (photograph by José J. Lunazzi).*

from the intention of the whole scene and from the intention of the iconography of the period. Would it be related to inverted mirror images, or more likely a political satire, as is common in European paintings? These questions await more exploration.

I also want to point out the more important fact that upward reflections on a horizontal mirror surface put everything that is on the sky down under the ground. This effect happens with clouds or the moon reflected in calm surfaces of water. Some indigenous tribes of Brazil believe that the image of the moon under the lake means that the moon itself is there (Arguello 2002). Likewise, a stone mirror can give the impression of an open window, or rather a hole, to see the moon under the ground. Equally impressive is the view of clouds when a plane mirror is held horizontally under one's nose. This creates the feeling of walking over the clouds, as if one were walking on the sky. Such experiences have been performed in art and education, having already fascinated thousands of people at public exhibits in South America (Lunazzi 2009b). Although the sensation is not as clear with 20% reflectivity mirrors as it is with common vanity mirrors of today, the possibility of its discovery by the ancients is worth consideration.

3. TO OBSERVE CONVERGING IMAGES THAT APPEAR IN FRONT OF CONCAVE MIRRORS

A concave mirror, besides giving an enlarged virtual image of a subject located within the focal distance of the mirror, also gives a reduced real (convergent) inverted image of a subject that is farther from that distance. The focal distance is equal to half the radius of curvature of the mirror. If the object is far away (a few meters for the case of Olmec mirrors) the image appears floating at its focal distance (between 10 and 30 cm). If the object is at a radius of curvature of distance from the mirror, the image appears precisely at the same distance and without distortion due to enlargement differences between transversal and longitudinal magnifications. A minimal separation must be given between object and image because only small angles from the center of symmetry are allowed to obtain an image with minimal aberrations. I can imagine two possible applications:

The first would be to produce an image resembling the sculptures existing at the Teotihuacan archaeological site, that is, the head of a feathered serpent exiting from a mirror (Taube 1992: figure 22). The size of a serpent's head is compatible with the well-known trick of hiding an object from the observer, leaving its illuminated part facing a concave mirror; even though an observer perceives the image as real, the perceived object fades when he/she tries to grab it. It is not difficult to imagine such a system being mounted on a wall that receives the solar light while only the light from the object is allowed to enter the internal room to impinge on the mirror, so that (besides very little residual entering light) only the image is visible. The previous up-down inversion of the serpent head compensates for the image inversion generated from the mirror. Around the image, real feathers could be located.

Another interesting possibility is the concave mirror being aimed at the moon. What a person holding a mirror can see when it is illuminated by the moon is a white disk floating in front that can easily represent and be associated with the moon, even if its reflectivity is only 20%. Lunazzi (1996) shows an image of a hand being reflected in an Olmec mirror. When in the presence of the mirror exhibited at the National Museum of Archaeology in Mexico City in 1995, or by stereo photography, a hand can be seen in front of the mirror.

4. TO IMPRESS PEOPLE WITH A BRIGHT REFLECTION OF THE SUN AS A SYMBOL OF THE DIVINE

Carlson (Carlson 1981, Figure 35) shows us the drawing of a god having a mirror on its forehead. The mirror could be reflecting the sun to give the impression not only of a bright head but of his association with the sun god. Due to the high intensity of the sunlight, 20% reflection is enough power to show the brightness on the mirror to a large audience. In order to distribute the reflected solar brightness to a horizontally located audience, more divergence must be given horizontally than vertically. Although a concave mirror makes convergent images at its focus, the spreading of light happens after, and a divergent beam is launched after the image. What the distant observer sees is a bright image of the sun in front of the mirror, but if he/she is many meters distant, the brightness may seem to be on the mirror itself. Plane mirrors can give bright sun reflections within a very limited angle of 0.5°. To achieve a larger horizontal distribution, a shorter focal distance must be given to the mirror by reducing its curvature on a horizontal plane. This is the case with some double cylindrical Olmec mirrors (Carlson 1981), improperly named as "parabolics," whose horizontal radius of curvature is shorter than the vertical radius. Due to their nonspherical shape, mirrors of this kind cannot usually produce good quality images, but brightness is more important than image quality in many cases. For the case of a specific jade mirror, it has been said: "Although Carlson (1981: 249) interprets this item as a 'non-functional' copy of an iron ore mirror, it is quite possible that, as with the metallic ore examples, this pectoral could have also been used in divinatory scrying. In such divination, the play of light is at least as important as the quality of the reflection" (Taube 2004: 142).

5. TO MAKE FIRE

By converging the sunlight to a focused image, it is possible to make fire with low-reflectivity mirrors if the appropriate material is prepared. In 1995, I conduced an experiment by equating the energy relationship of an 80% reflectivity mirror to reduce its diameter in such a way that the focused image had the energy equivalent to a 20% reflection mirror (Lunazzi 2011). The experiment was carried out twice, once in Campinas, Brazil, and the other time in Mexico City, at about 24°C and with some wind in both cases. It was in September, at the latitudes of the Tropics of Capricorn or Cancer, so that the sun's intensity for those latitudes was not maximal. My results can be compared to the case of Olmec mirrors using a 20% reflectivity spherical mirror of 42 mm in diameter. Although naturally dried leaves from native trees rendered only smoke (no attempt was made to produce flames by breathing on them, as is the usual indigenous procedure when making fire by wood and straw friction), flames soon were achieved by employing a cotton material that had previously been externally burned (Lunazzi 1995; see also de la Vega 2011). This experiment gives support to the idea that Garcilaso de la Vega (2011) was telling the truth when he described the Inca tradition of making fire by means of mirrors (Nordenskiöld 1926).

6. TO COMMUNICATE BY MEANS OF SUNLIGHT

At which distance could people communicate by using ordinary means, like sound or smoke signals? Can those signals be hidden from the enemy? Indeed, sunlight reflection on stone mirrors can be an advantage to other means if the day is sunny. It does not really matter if the mirror reflects 80% like the best aluminum or silver mirrors common nowadays, 55% like pyrite, or only 20% like hematite. Besides having access to bright sunlight that can reach the mirror, the most important thing is that the mirror is located properly to direct the light signal in a straight line during most of the day. At midday in tropical places the sun comes from above and can be reflected all around. At sunrise it is not possible to reflect the sunlight in a direction close to its incidence direction. Thus, not much more than three hours can be allowed for central transmission to any direction, but almost six hours could be allowed to reflect toward one position located at east or west, to a much longer distance than sound can propagate through mountainous or hilly terrain.

My research team and I demonstrated this by making two-directional communication at a distance of 1.4 km, a distance at which it is not possible for the naked eye to see the movement of a man's arms. We demonstrated this so easily that it makes it possible to imagine reaching similar results at much further distances. We employed ordinary silver mirrors with 20% average reflectivity obtained by attenuation of its normal reflectivity by means of dense black smoke deposition. The size of the mirrors was 15 cm wide and 18 cm high, a size which is close to the surface of the many plane stone mirrors being exhibited at the Museo del Oro at Lima, Peru (Lunazzi 2007), such as those shown in figure 6.4. To keep the mirrors in position, a base box filled with sand 12 cm deep was held at 1.3 m over the ground employing a prismatic piece of wood for the rough first angular zenithal positioning (see figure 6.5). The rough azimuthal positioning was obtained by rotating the box on its base while the fine angular adjustments were achieved by manually tilting the mirror. The crucial element is to aim the light toward its goal: a white piece of paper 20 cm large and 10 cm high fixed at a 4 m distance from the mirror on a support whose position can be easily regulated vertically and horizontally. Although we employed a photographic tripod, we know that a simple stick nailed on the

FIGURE 6.4. *A set of archaeological stone mirrors exhibited at Museo del Oro, Lima, Peru (photograph by José J. Lunazzi).*

ground could accomplish the same task. We watched from the center of the mirror (close and lateral to it) to see our target through a line close to the center of the white paper. Then we put the sunlight reflected by the mirror centered and over the white paper, leaving only the lower part of the light beam to illuminate the paper. At the other end what could be seen was a bright spot (figure 6.6) that the photograph or video (Lunazzi 2012) can hardly reproduce: the visual impression is much more effective. We could manually interrupt it with a piece of cardboard whose size was slightly larger than that of the mirror surface at a pace faster than twice a second, in order to code signals.

It is not difficult to characterize the situation by employing simple geometrical optics. The main value to be considered is the angular size of the sun, which is approximately 0.5°. This measurement determines the size of the angular illumination field, which is the same, so that the illuminated area at the aim is 1.5 m for each 150 m of distance. And the estimated visual and manual precision to locate the sunbeam on the white paper must always be better than that. We needed about 0.1° precision and this can be obtained to our aim with a sand base for the mirror. My team's calculations show that little departures from flatness or diffraction by the mirror's aperture provide negligible effects. As in the previous case of impressing an audience, again

FIGURE 6.5. *Setup employed for reflecting the sunlight, lateral view (photograph by José J. Lunazzi).*

FIGURE 6.6. *The bright spot obtained for communication, almost at the spiral center (photograph by José J. Lunazzi).*

image quality is not a fundamental concern (Taube 2004: 142). Even reflectivity power is not important, because just as sunlight is reflected by the lateral glass of a car's window (what corresponds to less than 5% reflectivity), light was also noticed at our observation base at 1.2 km distance from the car. It is clear that the communication distance can be further increased if we consider pyrite mirrors like the archaeological examples reported as belonging to the ancient Maya, because they have 60% reflectivity. Future experiments can shed light on the capacities of ancient Maya mirror specimens.

CONCLUSIONS

To situate the importance of the analysis for an archaeological audience, this chapter could work as a sort of exploratory or experimental think-piece. Experiments were made and necessary parameters analyzed to help in evaluating six possibilities for the ancient use of low-reflectivity stone mirrors already considered previously. Self-observation of one's face was reported, suggesting the need of sunlight. Observation of symmetrical images, although natural for plane mirrors, was applied to situations where the sky and its elements are important, and symmetry of the image was considered not only as a transversal property but longitudinal too, because depth is a very important element always present for the recognition of images (the reason for our binocular vision). Floating images, although evident when holding a concave mirror, were not specifically mentioned by other authors and are now proposed as a subject for future consideration. A physical explanation was given for why some mirrors were made with two curvatures. A possible way of making fire by using a mirror to manipulate sunlight was also detailed. And finally, a possibility of optical communication, not previously known to the author, was tested and analyzed. It is my hope that this work may help archaeologists in their evaluation of the fascinating civilizations of the ancient Americas, pre-Columbian cultures about which much remains to be revealed.

ACKNOWLEDGMENTS

The Pro-Reitoria de Extensão e Assuntos Comunitários-PREAC of State University of Campinas is acknowledged for funding that allowed for the purchase of some material. The Pro-Reitoria de Pesquisa of State University of Campinas is acknowledged for the "PIC Jr" fellowship given to students Cássia Sanches Delanhese, Caio Leonardo Duarte Bargas, and Rafael Pedro da Silva, who helped in performing and measuring the reflectivity adaptation of common mirrors. The Pro-Reitoria de Graduação of State University of Campinas is acknowledged for the SAE fellowship to student Alex Rafael da Costa and Henrique Guilherme Ferreira, who helped in performing the mirror experiments under the sun. Teroslau R. Perallis is acknowledged for his help in finding the proper elements for smoke deposition. The Museum of Archaeology of Mexico City, Mexico, the Museum Larco, Lima, Peru and the Museo del Oro, Lima, Peru are acknowledged for allowing me to take and reproduce pictures of museum pieces. Paul S. Baldi is acknowledged for grammar revision and Silvia L. Baldi for digital formatting of the photographs. The agreement between the Ministry of Education division CAPES and many journals allowed me to rapidly obtain important references. The open access to information on the Internet was a great source of help for finding other references, mainly the Google books service.

BIBLIOGRAPHY

Andersons, Miranda. 2007. *The Book of the Mirror: An Interdisciplinary Collection Exploring the Cultural Story of the Mirror*. England: Cambridge Scholars Publishing.

Arguello, Carlos A. 2002. *Personal Report*. Brazil: State University of Campinas.

Blainey, Marc G. 2007. "Surfaces and Beyond: The Political, Ideological, and Economic Significance of Ancient Maya Iron-Ore Mirrors." MA thesis, Department of Anthropology, Trent University, Peterborough, Ontario. Wayeb: The European Association of Mayanists. http://www.wayeb.org/download/theses/blainey_2007.pdf.

Borges, Jorge Luis. 1934. "Los Espejos Velados." *Critica, RMS* 2 (58): 7.

Calvo, María L., and Jay M. Enoch. 2007. "Ancient Peruvian Optics with Emphasis on Chavin and Moche Cultures." In Atti Della "Fondazione Giorgio Ronchi" Anno LXII. http://pendientedemigracion.ucm.es/info/giboucm/Download/Calvo_Atti_2007.pdf.

Carlson, J. B. 1981. "Olmec Concave Iron Ore Mirrors: The Aesthetic of a Lithic Technology and the Lord of the Mirror." In *The Olmec and Their Neighbors: Essays in Memory of Matthew W. Stirling*, ed. Elizabeth P. Benson, 117–147. Washington, DC: Dumbarton Oaks, The University of Maryland.

Carlson, J. B. 1993. "The Jade Mirror, An Olmec Concave Jadeite Pendant." In *Precolumbian Jade: New Geological and Cultural Interpretations*, ed. Frederick W. Lange, 242–250. Salt Lake City: University of Utah Press.

Cocteau, Jean. 1950. *Orpheus*. France: Andre Paulve Film and Films du Palais Royal.

de la Vega, Garcilaso. 2011. "Comentarios Reales de los Incas, 1609–1617." Last modified November 13. http://es.wikisource.org/wiki/Comentarios_reales.

Enoch, Jay M. 1999. "First Known Lenses Originating in Egypt about 4600 Years Ago." *Documenta Ophthalmologica* 99 (3): 303–314. http://dx.doi.org/10.1023/A:1002747025372.

Enoch, Jay M. 2006. "History of Mirrors Dating Back 8000 Years: Historical Perspective." *Optometry and Vision Science* 83 (10): 775–781. http://dx.doi.org/10.1097/01.opx.0000237925.65901.co.

Healy, Paul F., and Marc G. Blainey. 2011. "Ancient Maya Mosaic Mirrors: Function, Symbolism, and Meaning." *Ancient Mesoamerica* 22 (2): 229–244. http://dx.doi.org/10.1017/S0956536111000241.

Heizer, Robert F. 1967. Dumbarton Oaks Conference on the Olmec, October 28 and 29, 1967, edited by Elizabeth P. Benson. Washington, DC: Dumbarton Oaks.

Larsen, Gunnar. 1979. *Diagenesis in Sediments and Sedimentary Rocks,* Developments in Sedimentology, 25a. vol. 1. New York: Elsevier Publishing Company.

Lunazzi, J. J. 1995. "On the Quality and Utilization of Olmec Mirrors." Paper presented at the 2nd Reunión Iberoamericana de Óptica, Guanajuato, Mexico.

Lunazzi, J. J. 1996. "Olmec Mirrors: An Example of Archaeological American Mirrors." In *Trends in Optics*, vol. 3. ed. Anna Consortini, 411–421. Italy: International Commission for Optics—ICO, Ac. Press. http://dx.doi.org/10.1016/B978-012186030-1/50024-2.

Lunazzi, J. J. 2007. "Optica Precolombina Del Peru." Revista Cubana de Física 24 (2): 170–174. http://www.fisica.uh.cu/biblioteca/revcubfi/2007/vol24-No.2/rcf-2422007-170.PDF.

Lunazzi, J. J. 2009a. "Espelho Olmeca Convexo 1995." Last modified January 26. http://www.youtube.com/watch?v=IBabZ2AJAbQ.

Lunazzi, J. J. 2009b. "La Nube: A Maneira Mais Emocionante de se Experimentar Espelhos Planos." Caderno Brasileiro de Ensino de Física 26 (2): 416–425. http://www.periodicos.ufsc.br/index.php/fisica/article/view/11357/10891.

Lunazzi, J. J. 2011. "Haciendo fuego con espejo que simula un espejo arqueologico (Olmeca)." Last modified December 21. http://www.youtube.com/watch?v=KsRRFwibGZk&feature=youtu.be.

Lunazzi, J. J. 2012. "Projeto Comunicações Ópticas pelo Sol Simulando Espelhos Arqueológicos." Last modified January 23. http://youtu.be/YMlPEkVotOs.

Muelle, Jorge C. 1940. "Espejos precolombinos del Peru." *Revista del Museo Nacional. Museo Nacional (Peru)* 9 (1): 5–12.

Nordenskiöld, E. 1926. "Mirroirs convexes et concaves en Amérique." *Journal de la Société des Americanistes de Paris* 18 (1): 103–110. http://dx.doi.org/10.3406/jsa.1926.3607.

Olivier, Guilhem. 2004. *Tezcatlipoca: Burlas y metamorfosis de un dios azteca*. Mexico: Fondo de Cultura Económica.

Olson, Valerie F. 1984. "Olmec Magnetite Mirror: Optical, Physical, and Chemical Characteristics." *Applied Optics* 23 (24): 4471–4476. http://dx.doi.org/10.1364/AO.23.004471.

Rebollo, M. A. 1988. *Personal Report*. Com. Nac. de Energia Atómica. Argentina: CONEA.

Saunders, N. J. 1988. "Chatoyer, Anthropological Reflections on Archaeological Mirrors." In *Recent Studies in Pre-Columbian Archaeology*, edited by N. J. Saunders, O. de Montmollin, vol. 1: 1–39. BAR International Series 313. Oxford: British Archaeological Reports.

Simms, D. L. 1977. "Archimedes and the Burning Mirrors of Syracuse." *Technology and Culture* 18 (1): 1–24.

Taube, Karl A. 1992. "The Iconography of Mirrors at Teotihuacan." In *Art, Ideology, and the City of Teotihuacan*, ed. Janet Berlo, 169–204. Washington, DC: Dumbarton Oaks Research Library and Collection.

Taube, Karl A. 2004. *Olmec Art at Dumbarton Oaks*. Washington, DC: Dumbarton Oaks; http://www.doaks.org/resources/publications/books-in-print/pre-columbian-art-at-dumbarton-oaks/olmec-art-at-dumbarton-oaks.

Xiu, Yi. 1996. "Suna fajrilo antau 3000 jaroj." *Rev. El Popola Chinio* 5:48.

7

The 49 objects of iron pyrite (46 ornaments and three plaques) discussed in this chapter came from excavations in three Middle Formative cemeteries in the valley of Mascota, Jalisco, one cemetery radiocarbon dated to approximately 1000 BC and the other two to approximately 800 BC. Only 17 other objects of iron pyrite (thought to have been plaques of a mosaic) have been recovered from Middle Formative contexts in West Mexico, and those are from tomb fill at El Opeño, Michoacan (Oliveros 2004: 148). Therefore, these 49 objects from Mascota constitute by far the largest of the only two collections of Middle Formative iron-pyrite objects recovered from professional excavations in West Mexico. A major significance of these objects is that they link the Mascota valley into a Mesoamerican system of exchange in somewhat exotic stones such as iron pyrite, hematite, magnetite, and quartz that have special reflective qualities, along with jadeite and some other stones important primarily for their green or red color. This system of exchange included the Basin of Mexico, Valley of Oaxaca, the Olmec heartland in Veracruz and Tabasco, and reached as far as the central highlands and southern coast of Guatemala. This was not only a system of exchange of exotic objects; the objects themselves were associated with the expansion to the northwest of agriculturalists with a Mesoamerican ideology that emphasized the importance of the sun, water, and fertility.

Iron Pyrite Ornaments from Middle Formative Contexts in the Mascota Valley of Jalisco, Mexico

Description, Mesoamerican Relationships, and Probable Symbolic Significance

Joseph B. Mountjoy

DOI: 10.5876/9781607324089.c007

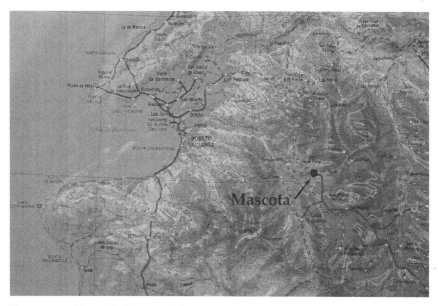

FIGURE 7.1. *Map of the location of the town and valley of Mascota in the State of Jalisco in West Mexico.*

BACKGROUND

During a five-year project (2001–2005), supported by the National Geographic Society and the Foundation for the Advancement of Mesoamerican Studies, Inc., we investigated Middle Formative society in the Mascota valley of Jalisco in West Mexico (figure 7.1). The Mascota valley was seen to be a particularly important location to study Middle Formative adaptation to the West Mexican geographical area because it appears to have been a center of Middle Formative development in Jalisco, and indeed in West Mexico as a whole. We have located 12 sites with Middle Formative remains in the Mascota valley, based on burial offerings found at each of the sites. These 12 sites account for approximately 25% of the total number of sites with pre-Hispanic remains so far registered in the Mascota valley. By comparison, I do not know of 12 other Middle Formative sites thus far registered in the entire remainder of the State of Jalisco.

In Nayarit there are only two known Middle Formative sites, both of which form part of a habitation area near the northern base of La Contaduria hill on the northern edge of San Blas, on the central coast (Mountjoy and Claassen 2005). This habitation is characterized by low house mounds on artificial terraces, associated with abundant refuse of the coastal shellfish, crabs, fish, and birds that were heavily relied upon by these people for their subsistence.

In far western Michoacan there is one Middle Formative site, the cemetery of El Opeño, where several staircase and chamber tombs have been excavated and their contents studied (Oliveros 2004). Capacha, a Middle Formative archaeological phase in Colima, was originally defined by Isabel Kelly (1980) on the basis of her study of offerings from 10 cemetery sites, one of which was over the western border of Colima into Jalisco.

What is particularly important about the remains found in these scarce Middle Formative sites is what they reveal about the northwestern spread of Neolithic-type small village farmers out of Southern and Central Mexico into areas of West Mexico that have little or no evidence of a preceding Archaic-type hunting-and-gathering adaptation. It appears that due in part to the extremely harsh and lengthy dry season, in most areas of West Mexico habitation was not possible without possession of cultivated plants such as maize, squash, and especially beans. The Mascota valley was an especially attractive environment for these early garden farmers, due to the extensive humid and fertile soil of the old lake basin that constitutes the valley floor.

Our investigations in the Mascota valley were focused on five problems related to understanding the development of the Middle Formative period (ca. 1000–700 BC): (1) the ecological/cultural adaptation of these Middle Formative people who were the first sedentary agriculturalists to colonize this part of West Mexico (Mountjoy 1974; Kelly 1980); (2) the relationship of their remains with the Olmec archaeological culture (Noguera 1942); (3) the relation of these remains in the Mascota valley with non-Olmec remains in the Central Mexican highlands, especially those recovered from the site of Tlatilco (Noguera 1942; Mountjoy 1970; Oliveros 1974); (4) the possible long-distance connections between the Middle Formative remains in the Mascota valley and Andean South America, specifically southern Ecuador (Mountjoy 1974; Kelly 1980); and (5) the Middle Formative roots of the Shaft Tomb Mortuary Complex of the Late Formative (ca. 300 BC to AD 300) in West Mexico (Kelly 1980).

In the course of our investigations we located and studied 12 Middle Formative archaeological sites and conducted extensive excavations at three of them: Los Coamajales (MA–20); El Pantano (MA–9); and El Embocadero II (MA–19). After five years of fieldwork and analysis some provisional answers can be provided for the five research problems (Mountjoy 2012).

For the first one (1), the ecological/cultural adaptation, we found the habitation sites concentrated along the piedmont overlooking the fertile and humid lowlands of an ancient lake basin. The settlements appear to have been small villages that probably were abandoned after 50 years or so, probably because of decline in the agricultural productivity of the soil. The people subsisted on the domesticated crops of maize, beans, and squash, supplemented by hunting of deer and turtles, as well as the collecting of wild plants, some of which were probably used to alleviate the pain of dental caries and dental abscesses from which a high percentage of the adult population suffered. They had cemeteries that constituted a "community of the dead" in which they placed the deceased whose remains had been curated during the year until they were buried during an annual community ceremony that probably coincided with the transition from the dry season to the rainy season and thus symbolized community/ population renewal linked to the growing season.

In terms of the second problem (2), ties to the Olmec archaeological culture of Southeastern Mexico, although admittedly remote, were apparent in the growling jaguar effigy vessel found in one of the shaft-and-chamber tombs at Los Coamajales, as well as in the jewelry that included jadeite from the same Guatemalan source used by the Olmecs.

As for the third (3), relations with non-Olmec archaeological cultures of the Central Mexican highlands were more evident, consisting of the same forms of pottery vessels often referred to as the "West Mexico Component" at Tlatilco in the Valley of Mexico and similar sites in the State of Morelos. These included such items as composite silhouette "water cover" and stirrup spout vessel forms, as well as large solid figurines with traits related to the central highland deities of Xipe Totec, Quetzalcoatl, and Mother Creation.

Regarding the fourth problem (4), ties with South America, specifically Ecuador, were indicated by such pottery items as a two-tube-and-bridge vessel, stirrup-spout vessels, phallic-form bottles, and head rests.

As for the fifth problem (5), the roots of the Shaft Tomb Mortuary Complex of the Late Formative were evident in such things as the oldest shaft-and-chamber tombs thus far discovered in Jalisco, an emphasis on pottery vessels of gourd form, a variety of large solid pottery figurines that included male and female "original pairs" and female mourners, as well as hollow dog-effigy vessels, some of which bear human faces or masks.

In general, the Middle Formative sites in the Mascota valley reveal a settlement pattern of hamlets to small villages focused on occupying areas just upslope around the edge of the old lake basin. The burial offerings indicate cultural contacts with areas as distant as southern Nayarit, central Colima, the Central Mexican highlands, the Motagua valley of northern Guatemala, and southern Ecuador (Mountjoy 2012). Some of these contacts must have been

TABLE 7.1. Radiocarbon dates from excavated contexts

Context	Normal Date	Calibrated 95%	Laboratory
LOS COAMAJALES (MA-20)			
Unit #2, "Capacha" urn	890 BC ± 40	1,100–900 BC	(Beta 202346)
Pit #1, "Capacha" urn	730 BC ± 40	900–800 BC	(Beta 211717)
EL PANTANO (MA-9)			
Pit #1, tomb entrance	690 BC ± 40	840–790 BC	(Beta 154581)
Pit #6, 91–158 cm	740 BC ± 40	910–800 BC	(Beta 159351)
Pit #9, 125 cm south burial	540 BC ± 40	790–420 BC	(Beta 154582)
Pit #9, 100 cm fill	390 BC ± 50	720–620 BC	(Beta 154608)
		590–390 BC	
Pit #10, burial level #2	790 BC ± 40	1,000–820 BC	(Beta 159353)
Pit #10, burial levels #2 and #3	740 BC ± 40	920–800 BC	(Beta 159354)
EL EMBOCADERO II (MA-19)			
Pit #12, olla interior	650 BC ± 40	820–770 BC	(Beta 211716)

via watercraft along the Pacific coast and probably involved the importation of foreign jewelry that to the local farmers was of great symbolic value relating to wet season fertility, in exchange for local exotic mountain highland plants with analgesic, hallucinogenic, or presumed curative properties (Mountjoy 2012).

The information resulting from these investigations also allowed us for the first time to divide the Middle Formative occupation in West Mexico into an early phase (as seen at Los Coamajales) and a late phase (as seen at El Pantano and El Embocadero II), a division evident in the radiocarbon dates obtained from excavated contexts at these three sites (table 7.1).

PYRITE ORNAMENTS OF THE MASCOTA VALLEY, JALISCO

We recovered iron-pyrite ornaments from all three of the Middle Formative cemeteries where we conducted excavations (figure 7.2). These cemeteries, and thus the pyrite ornaments recovered from them, can be ordered chronologically from the oldest (Los Coamajales) to the latest (El Pantano and El Embocadero II), based in part on the radiocarbon dates presented in table 7.1. This dating places the iron-pyrite ornaments recovered from the three sites among the oldest such ornaments thus far discovered in Mesoamerica. A brief description of the sites and the associated material will be presented next.

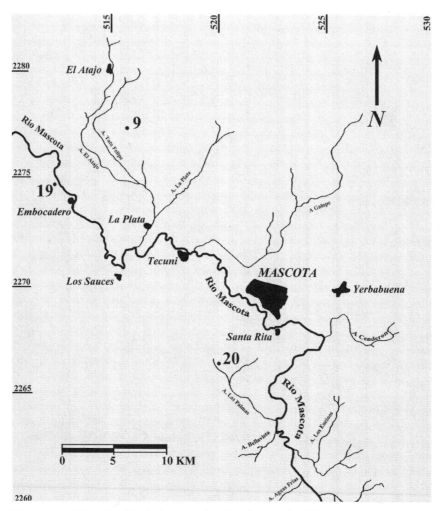

FIGURE 7.2. *Map showing the location of the Los Coamajales (MA–20), El Pantano (MA-9), and El Embarcadero II (MA–19) sites in the valley of Mascota, Jalisco.*

LOS COAMAJALES

This is a cemetery site located in the mountains that form the southern side of the Mascota valley, at 1640 m above sea level, and approximately 500 m above the valley floor (figure 7.2, site 20). The name *Los Coamajales* refers to the general area, but the local name of the precise place of the cemetery is Las Lomas due to the presence in this location of two low hills of soft volcanic

FIGURE 7.3. *Iron-pyrite ornament from Los Coamajales, broken by looters (photograph by Joseph B. Mountjoy).*

earth lacking stones that stand out in the midst of a pine forest strewn with volcanic rocks. The cemetery is not associated with any nearby habitation site. It appears that the remains buried here came from some habitation site probably at the southern foot of the mountain area where the deceased were cremated and their remains placed in urns to be buried in this special place high in the mountains.

The Los Coamajales cemetery had been heavily looted many years prior to our intervention, thus much of our initial work involved the screening of looters' spoil dirt. This effort yielded five fragments of what appears to be a very large ornament of some sort made of iron pyrite (figure 7.3). It may have been worn suspended on the chest but we lack the extremities that may have had a hole or groove for attachment. Three other items of iron pyrite were found in an unlooted shaft-and-chamber tomb at more than 2 m depth from the surface (figure 7.4). These were three small plaques (figure 7.5) polished on one side, which may have been set as ornaments in an object of wood. These plaques were associated with fragments of two unfired figurines, a small

FIGURE 7.4. *Shaft-and-chamber tomb at Los Coamajales where three small iron-pyrite plaques were recovered from the floor, with Nathan Mountjoy excavating (photograph by Joseph B. Mountjoy).*

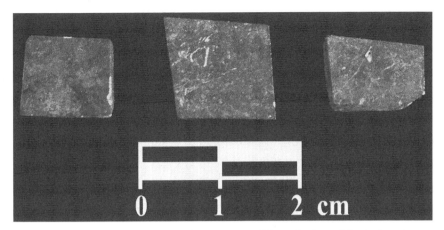

FIGURE 7.5. *Three small iron-pyrite plaques recovered from the floor of the shaft-and-chamber tomb pictured in figure 7.4 (photograph by Joseph B. Mountjoy).*

pottery cup, and an effigy vessel in the form of a jaguar with a Capacha-style growling mouth reminiscent of Olmec jaguar imagery. The two calibrated radiocarbon dates from a Capacha-associated context in the cemetery suggest placement somewhere in a range of 1100–800 BC, say around 1000 BC for a central date.

Arturo Oliveros (2004: 148, and plate 8) found similar small plaques of iron pyrite in fill during his excavations in a shaft-and-chamber tomb cemetery at El Opeño, Michoacan, and he assumed them to be part of a mosaic with wood backing. Oliveros obtained four radiocarbon dates for Tomb 7 at El Opeño, and a recent recalibration of those dates places them between 1200 and 1000 BC, overlapping with the dates from the Los Coamajales cemetery (Mountjoy 2012). Isabel Kelly (1980) did not recover any iron-pyrite ornaments with Capacha-phase associations in Colima.

EL PANTANO

The El Pantano site (figure 7.2, site 9) is located in the extreme northwest part of the Mascota valley at an elevation of approximately 1,260 m above sea level, and only 40 straight-line km from Puerto Vallarta on Banderas Bay to the west. The cemetery is located at the foot of a hillside elevated on a slight terrace along the northern edge of the ancient lake that once filled the Mascota valley. The cemetery seems to have been located on the eastern edge of a rather extensive Middle Formative village, but little remains of the Middle

Formative habitation on the surface because of the occupation and alteration of this site by people of subsequent pre-Hispanic archaeological cultures. The burials were placed deep enough to have for the most part avoided destruction by the later occupations except for occasional secondary burials or storage pits.

We excavated approximately 43 Middle Formative burial pits (many of "shaft-and-pit" form) in the El Pantano cemetery, recovering over 628 burial offerings, including 140 ornaments. Of these, a total of 42 were of iron pyrite. In only one instance is there a possible association of one of these ornaments with an infant. They were apparently acquired and worn by a relatively large number of both male and female adults in this society. Forty of these ornaments were pendants and two were beads. Several of the pendants have a pit in the center and a notch at each end, a form that suggests the intent was to portray an image of the sun with rays (figure 7.6). When polished, these ornaments would have had a reflective surface like polished silver. There are instances of association of these ornaments with crania in bundle burials, suggesting that they might have been worn on the forehead; but when these ornaments were found associated with articulated burials, this context suggests that they were suspended around the neck on the upper part of the person's chest.

The great majority of these iron-pyrite ornaments, unlike the four iron-pyrite ornaments from the Los Coamajales cemetery, are "doublets," with the outer face of iron pyrite and a backing of quartz. That this "doublet" appearance was the result of their natural process of formation was confirmed by Dr. James Luhr, Head of the Geology division of the Smithsonian Institution (personal communication, 2002), who sectioned three of these pieces and examined the sections under high-power magnification. There are, however, a few of the iron-pyrite ornaments that lack the quartz layer and are simply triangular plaques with a hole for suspension. In another case we found two small ball-shaped beads of iron pyrite that were found with the cranium of an extended and articulated burial of a man 25 to 35 years of age, and in a location associated with his mandible that suggests they were worn in the fashion of a necklace.

Six radiocarbon dates from the cemetery place these ornaments within a calibrated chronological range of approximately 1000–400 BC, with an average of the dates at 786 BC.

EL EMBOCADERO II

El Embocadero II cemetery is located in the southwestern part of the Mascota valley (figure 7.2, site 19) at an elevation of 1,270 m above sea level, on

FIGURE 7.6. *Iron-pyrite ornaments recovered from five different tombs in the El Pantano cemetery (photograph by Joseph B. Mountjoy).*

the other side of the valley from and almost directly south of the El Pantano site. The cemetery was first encountered during a project of road improvement for better access to farmlands, and then in 2004 the central part of the cemetery was ransacked by some professional looters from Tala, Jalisco. However, these looters left undisturbed burials, mostly shaft-and-chamber tombs, on the northern and southern sides of the area they pillaged. The Middle Formative habitation area was probably somewhere nearby, but erosion, which has removed something between 50 cm and a meter of the surface of this hillock on a high terrace on the southern side of the Mascota river, may have been responsible for the disappearance of any remains of Middle Formative habitation that were once left on the surface.

We excavated nine unlooted burial pits at this cemetery and cleaned out four looted burial pits. Most burials were deposited in relatively shallow shaft-and-chamber tombs. In this process, we recovered six ornaments of stone and three iron-pyrite "doublet" ornaments (figure 7.7) from the cemetery, plus one ornament of black slate and two ornaments of *Spondylus* shell. One of the

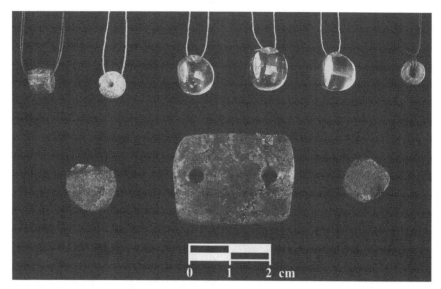

FIGURE 7.7. *Stone ornaments recovered from excavations at the Los Coamajales II cemetery: from left to right (top row) jadeite, possible amazonite, cabochoned and faceted quartz (x3), and jadeite; (bottom row) iron pyrite (x3) (photograph by Joseph B. Mountjoy).*

iron-pyrite ornaments was found in the cultivated zone, probably discarded from the looting of one of the burials at the site, but the other two were found near the mandible of an extended and articulated burial of a male 45 to 55 years old, suggesting they were worn around his neck as on a necklace.

This cemetery has one calibrated radiocarbon date of 820–770 BC, with a central date of 805 BC, and this date indicates that the El Embocadero II cemetery is roughly contemporary with the El Pantano cemetery, so the similarity between the iron-pyrite ornaments at both sites is not surprising.

FURTHER OBSERVATIONS

Iron-pyrite "doublets" of the kind found at El Pantano and El Embocadero II were not found by Oliveros (2004) in the shaft-and-chamber tombs at El Opeño in Michoacan, nor were they found by Kelly (1980) in the pit burials of the Capacha phase in Colima. Farther afield, such objects do not seem to be present at the roughly contemporary site of Tlatilco in the Basin of Mexico (García Moll et al. 1991), despite many other similarities in ornaments, pottery, and stone artifacts between El Pantano and Tlatilco. Nonetheless, García Moll (1999: 21) does cite the discovery of some offerings of iron pyrites in the burials at Tlatilco, but they are apparently not ornaments.

There have been "mirrors" or "reflectors" of different iron ores found at the Olmec centers of San Lorenzo in Veracruz and La Venta in Tabasco (Pires-Ferreira 1976), as well as Tlatilco in the Basin of Mexico (García Moll 1999: 21). At San Lorenzo they are dated within the range of 1000–750 BC (Pires-Ferreira 1976: 324), and at La Venta to Middle Formative constructional phases (Pires-Ferreira 1976: 317), presumably in the range of 900–500 BC (Diehl 2004: 60–61).

None of these "mirrors" however is of iron pyrite; most are of hematite probably obtained from the Cerro Prieto source near Tehuantepec, Oaxaca (Pires-Ferreira 1976: 322), although two are of magnetite and this links them to a workshop at San José Mogote, Oaxaca, where such "mirrors" appear to have been produced in abundance, roughly in the time span of 900–750 BC (Pires-Ferreira 1976: 324).

Even though the iron ore used and the form produced for the "mirrors" found at the Olmec centers in Veracruz and Tabasco is different from the ornaments found in the Mascota valley of Jalisco, there is no direct link between the magnetite ornaments produced at San José Mogote, Oaxaca, and those of iron pyrite found in the valley of Mascota, Jalisco, although the time period for all is roughly the same. However, at least some of the "mirrors" found at the San José Mogote, Oaxaca, workshops (Pires-Ferreira 1976: 319, Figure 10.11) are of shapes similar to some of those found at El Pantano and El Embocadero II, and in all instances it is believed that such ornaments were worn on the chest for the purpose of reflecting sunlight (Pires-Ferreira 1976: 324). This clearly would seem to link the Mascota valley of Jalisco to the general Mesoamerican system of what Pires-Ferreira (1976) calls "interregional exchange networks" during the Middle Formative, as is also the case with the jadeite ornaments from Middle Formative contexts in the Mascota valley, and presumably the quartz ornaments as well.

There is no evidence that the 49 iron-pyrite ornaments (at Los Coamajales, El Pantano, and El Embocadero II) nor the 45 of other stone, primarily jadeite (at El Pantano and El Embocadero II) or the 13 of quartz (at Los Coamajales, El Pantano, and El Embocadero II) were fabricated locally.

One of the jadeite beads was inspected by Sorena Sorenson of the Department of Geology at the Smithsonian Institution and the material appears to have come from the known source in the Motagua river area of Guatemala, based on the abundance of the mineral albite (personal communication, 2002). It appears all of the jadeite ornaments found at El Pantano and El Embocadero II have the same origin.

Some of the quartz ornaments are of exquisitely fine workmanship, including two trumpet-shaped pendants from Los Coamajales (Mountjoy 2006) that have counterparts in a similar piece attributed to the Capacha phase in Colima (Kelly 1980: 82, Figure 35e) and another trumpet-shaped quartz pendant at the depth of 200–250 cm in a stratigraphic pit at the Rancho La Pintada site in the Tomatlán valley of Jalisco (Mountjoy 1982b: 159), an ornament that probably pertains to the Capacha-phase occupation of the Tomatlán valley.

Even more sophisticated are the double faceted and cabochoned pendants from El Pantano (site # 1) and El Embocadero II (site # 3) (figure 7.7). These may be the oldest gemstones in the world having facets made with the intent of reflecting light back out of the front of the stone. The only other faceted quartz ornaments known in Mesoamerica are two turtle-shaped ornaments from the Olmec site of La Venta in Tabasco (Drucker et al. 1959: 149–150, plate 28). These pieces in a form suggesting a turtle shell have three facets on the underside apparently for the purpose of reflecting the light out of the curved front if the pieces were worn as pendants or attached to clothing. Oliveros found six faceted beads at El Opeño, Michoacan (Oliveros 2004: Lámina 8, center, left; Robles Camacho and Oliveros Morales 2007: 9, Figure 2a), but the faceting there appears to have been done to create a distinctive shape since all the pieces are of opaque jadeite.

The jadeite used for ornaments, the high quality of the craftsmanship of the quartz ornaments, and the labor involved in the production of the ornaments of iron pyrite suggest importation of all three types of items to Middle Formative communities in the Mascota valley of Jalisco. These ornaments must have been produced by highly skilled lapidaries at one or more workshops that were probably located near the source of the raw material. Since we know the jadeite came from the Motagua, Guatemala, source, it seems logical to suggest that the quartz and iron-pyrite ornaments were likewise crafted there.

These exotic items were probably highly valuable pieces in Pacific coastal trade that in the case of the Mascota valley sites involved exchange of exotic local plants with medicinal or psychotropic properties for these ornaments manufactured in faraway Guatemala. Nearly all such stone ornaments are associated with adults, both male and female, persons who could acquire the local products for exchange by collecting them in the mountains surrounding the Mascota valley, at an elevation of around 1,600 m or more above sea level, at least 500 m above the valley floor.

But other than the beauty and novelty of these ornaments made of exotic materials, they probably had an important symbolic value based on color and other characteristics. They do not seem to have been markers of wealth or status. For the Neolithic-type Middle Formative farmers who colonized the Mascota valley and much of the rest of West Mexico there seem to be three important factors in their survival: sun, water, and fertility. The central importance of these three factors for rainfall farmers in West Mexico carries on through the Spanish Conquest and even up into modern times. Archaeologically, this sun-water-fertility theme is expressed as well in some forms of pottery (including bottles in water gourd shape or phallic shape), and in certain pottery decorations (such as the female pubic triangle decoration found on some of the early Capacha burial urns), as well as most of the designs executed in rock art that remain as evidence of rituals of renewal associated with the crucial transition from the dry season to the rainy season (Mountjoy 1982a, 2001).

As for the Middle Formative ornaments in general, the jadeite and other green or blue-green colored stones (figure 7.7) were likely associated with the greening of the landscape and their crops during the rainy season. Leonardo López Lujan (2009) has referred to jadeíta as "petrified water" in relation to its use for offerings at the Great Temple of Tenochitlan, and cites the *Florentine Codex* as registering that greenstones were believed by the Mexica to have the double property of attracting and exuding humidity, thereby making them excellent symbols of fecundity (López Lujan 2009: 54). The quartz (figure 7.7), due to its clarity and transparency, may also have had a direct association with water, but the use of some quartz crystal pendants and the shaping of faceted quartz stones appear to emphasize in these cases the reflection of the rays of the sun.

The iron pyrite pendants (Figures 7.6 and 7.7) had a silvery surface that likewise would have made them excellent reflectors of the sun's rays and thereby also indirectly related to the onset of the rainy season that results in the greening of the vegetation and the growing of the crops.

The succinct way of summarizing the relation of the iron-pyrite ornaments, and indeed all stone jewelry of this Middle Formative development in the Mascota valley of Jalisco, Mexico, to the Middle Formative culture as a whole, is to put them into the context of "rituals of renewal" (Mountjoy 2001) that were carried out by these small-village, agricultural peoples who were ritually focused on the critical transition from the dry season of scarcity to the wet season of abundance. Crucial to this transition was the deity of the sun, which would cease to burn the dead, dry, and sterile earth and begin to release the life-giving rainfall of the wet season. Burial rites at El Pantano and El Embocadero appear to have been conducted at the transition between the dry and wet seasons, and to have involved interment of all the deceased for the previous year, whose remains had been curated in a special place until the annual ceremony. The inhabitants wore stone jewelry that generally signified water and fertility by having special properties to reflect the sun's rays or being of a clear transparent or green color that signified water and the wet season vegetation.

BIBLIOGRAPHY

Diehl, Richard A. 2004. *The Olmecs: America's First Civilization.* Ancient Peoples and Places Series. London: Thames and Hudson.

Drucker, Philip, R. F. Heizer, and R. J. Squier. 1959. *Excavations at La Venta Tabasco, 1955.* Smithsonian Institution Bureau of American Ethnology Bulletin 170. Washington, DC: United States Government Printing Office.

García Moll, Roberto, D. Juárez C., C. Pijoan A., and María E. Salas C. 1991. *Catálogo de entierros de San Luís Tlatilco, México.* Serie Antropología Física-Arqueología. México City: INAH.

García Moll, Roberto. 1999. "Tlatilco: Prácticas funerarias." *Arqueología Mexicana* 7 (40): 20–23.

Kelly, Isabel T. 1980. *Ceramic Sequence in Colima: Capacha, an Early Phase.* Anthropological Papers of the University of Arizona no. 37. Tucson: University of Arizona Press.

López Lujan, Leonardo. 2009. "Aguas petrificadas: las ofrendas a Tláloc enterradas en el Templo Mayor de Tenochtitlán." *Arqueología Mexicana* 16 (96): 52–57.

Mountjoy, Joseph B. 1970. "Prehispanic Culture History and Cultural Contact on the Southern Coast of Nayarit, Mexico." PhD diss., Southern Illinois University, Carbondale, IL.

Mountjoy, Joseph B. 1974. "San Blas Complex Ecology." In *The Archaeology of West Mexico,* ed. Betty Bell, 106–119. Guadalajara: Sociedad de Estudios Avanzados del Occidente de México.

Mountjoy, Joseph B. 1982a. "An Interpretation of the Pictographs at La Peña Pintada (Jalisco, Mexico)." *American Antiquity* 47 (1): 110–126. http://dx.doi.org/10.2307 /280057.

Mountjoy, Joseph B. 1982b. "Proyecto Tomatlán de Salvamento Arqueológico: Fondo etnohistórico y arqueológico, desarrollo del Proyecto, estudios de la superficie." *Colección Científica: Arqueología* 122. México City: INAH.

Mountjoy, Joseph B. 2001. "Ritos de Renovación en los petroglifos de Jalisco." *Arqueología Mexicana* 47:56–63.

Mountjoy, Joseph B. 2006. "Excavation of Two Middle Formative Cemeteries in the Mascota Valley of Jalisco, México." Last modified February 15. www.famsi.org /reports/03009.

Mountjoy, Joseph B. 2012. *El Pantano y otros sitios del Formativo Medio en el valle de Mascota, Jalisco*. Guadalajara: Secretaría de Cultura de Jalisco, Universidad de Guadalajara Centro Universitario de la Costa, Ayuntamiento de Mascota, Acento Editores.

Mountjoy, Joseph B., and Cheryl P. Claassen. 2005. "Middle Formative Diet and Seasonality on the Central Coast of Nayarit, Mexico." In *Archaeology Without Limits*, ed. B. P. Dillon and M. A. Boxt, 267–282. Lancaster, CA: Labyrinthos Press.

Noguera, Eduardo. 1942. "Exploraciones en El Opeño, Michoacán." In *XXVII Congreso Internacional de Americanistas*, vol. 1: 574–586. México: INAH.

Oliveros, J. Arturo. 1974. "Nuevas exploraciones en El Opeño, Michoacán." In *The Archaeology of West Mexico*, edited by Betty Bell, 182–201. Guadalajara: Sociedad de Estudios Avanzados del Occidente de México.

Oliveros, J. Arturo. 2004. *Hacedores de tumbas en El Opeño, Jacona, Michoacán*. Michoacan: El Colegio de Michoacán and the Ayuntamiento de Jacona.

Pires-Ferreira, Jane W. 1976. "Shell and Iron-Ore Mirror Exchange in Formative Mesoamerica, with Comments on Other Commodities." In *The Early Mesoamerican Village*, ed. Kent V. Flannery, 311–328. New York: Academic Press.

Robles Camacho, Jacinto, and Arturo Oliveros Morales. 2007. "Estudio mineralógico de lapidaria prehispánica de El Opeño, Michoacán: Evidencias de organización social hacia el Formativo medio en el occidente de México." *Arqueología* 35: 6–22.

8

Pre-Hispanic Iron-Ore Mirrors and Mosaics from Zacatecas

Achim Lelgemann

The first Mesoamerican farmers moved into the inter-montane valleys and eastern piedmont regions of the Sierra Madre Occidental in what are now the intricate borderlands of Zacatecas and Jalisco during the Late Formative and Early Classic periods, ca. AD 100–300 (Lelgemann 2009: 130–131). There they found a wide array of mineral resources that they turned into pigments and diverse objects of personal adornment or ceremonial use. The mountain ranges of the Western Sierra Madre are basically of volcanic origin and formed in several violent eruptive events during the Late Tertiary and Early Quaternary periods. These ranges are made up of immense accumulations of extrusive igneous rock that constitute the matrix of all sorts of iron ores like hematite and its lustrous variety, specularite, as well as ilmenite, magnetite, pyrite, and chalcopyrite (*Monografía geológico-minera del estado de Zacatecas* 1991: 17–30; *Síntesis Geográfica de Zacatecas* 1981: 19–23).

Conglomerates of all these minerals are readily available at or near the surface, but must have been obtained in even larger quantities through the extensive mining operations the region has been famous for from pre-Hispanic to modern times (Weigand 1982). Exploitation of mineral-rich deposits was carried out in shallow superficial pits and beds of natural watercourses but also, in different scales of intensity and depth, in subterranean shaft-and-tunnel systems, reaching gigantic proportions in the Suchil Branch of the Chalchihuites culture during the Middle and Late

DOI: 10.5876/9781607324089.c008

Classic from around AD 500 to 900 (Weigand 1968; Schiavitti 1996). The extraction, processing, and artifactual transformation of the target minerals were of enormous economic importance, second only or equal to agriculture. They might even have been prime motives of attraction that lured Mesoamerican settlers from West and Central Mexico into the inhospitable environments of the Sierra Madre, where farming based on permanently erratic rainfalls has always been and still is a highly risky enterprise (figure 8.1).

This chapter summarizes what is currently known about the pre-Hispanic use of native iron ores, mostly pyrite and specular hematite. We focus on the manufacture of mirrors and mosaics within the confines of the Mesoamerican occupation zone in southern and western Zacatecas during the Classic and Postclassic epochs, from the beginning of the Christian era to the Spanish conquest of the 1530–1540s. The analysis draws largely upon recent finds of both unprocessed raw materials and finished artifacts in the context of mortuary deposits at two major sites in the state of Zacatecas. Here, high-status individuals were accompanied by rich and varied grave goods including ornamental or ceremonial items made of small iron-ore plates glued to slate disks or perishable carrier materials.

Beyond the merely descriptive presentation of the physical evidence available at this point, the objects themselves and especially their contexts and associations have far-reaching interpretive implications for our understanding of the acquisition, maintenance, and ostentation of political power in Middle and Late Classic chiefdoms of northwestern Mesoamerica. These will also be discussed in relation to very specific patterns of religious belief and their manifestation in elaborate ceremonialism, as they have been identified in major architectural compounds of monumental sites in Zacatecas like Alta Vista or La Quemada.

THE EVIDENCE

Tepizuasco

The earliest archaeological evidence for the artifactual use of iron ores in pre-Hispanic Zacatecas was retrieved at the mesa site of Tepizuasco, 8 km to the northeast of Jalpa in the central portion of the Juchipila drainage (Weigand et al. 1999). Although recovered from indistinct debris accumulations, the architectural and stratigraphic associations and contexts strongly indicate that the minute pieces date to the Apozol and Huanusco phases of the Late Formative and Early Classic periods (ca. AD 200–400). Occasional occurrences of specular hematite in indisputable anthropogenic transformations are largely

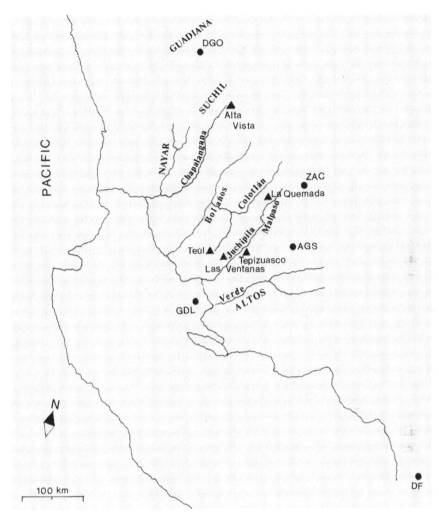

FIGURE 8.1. *Map of Northwestern Mesoamerica, showing major archaeological regions and sites mentioned in the text.*

represented by tiny *tesserae* 2–5 mm in length and less than 1 mm thin, normally in irregular shapes but generally with polished and striated surfaces. Although specularite (Fe_2O_3) already comes in lamellar aggregates, sometimes in rose or rosette-like conglomerates with a natural silvery shine (Schumann 1993: 104–105), human intervention in the creation and use of small chips or flakes is undoubtedly documented by the presence of deliberately created geometric contours different from naturally formed crystals. They occur as triangles and in one instance even a perfectly shaped diamond. In a few cases, the beveling of plaque edges can already be observed, a technique that was to become a standard procedure during Middle and Late Classic times. Indicative of mass processing of tesserae and the manufacture of composite artifacts like mosaics, the oblique grinding of rims is supposed to facilitate the insertion of small plaques into the soft mass of the vegetable substances used as glue. At the same time, this technique reduces the joints between individual pieces to an extremely fine line that is hardly perceptible.

At present it is still impossible to state whether these isolated finds mark the beginnings of a local industry of jewel production, perhaps in specialized workshops, based on the regionally available raw material, or if the few items were actually imports processed at other sites where craft specialization was already more advanced. Independent from these uncertainties, since the early finds at Tepizuasco are exclusively associated with high-status residential and ceremonial compounds on the flat mesa top, it is reasonable to infer that these artifacts are linked to the higher echelons of society during the early occupation of the site.

Although no shaft tombs have been identified there, these are known from nearby settlements like Las Ventanas 40 km to the south in the Juchipila drainage, Teúl in the neighboring Tlaltenango valley, and from several major sites of the Bolaños gorge further to the west. These provinces were all linked together in a larger interaction sphere, sharing architectural patterns, ceramic types, and figurine traditions (Lelgemann 2010). Within the Juchipila valley, Tepizuasco boasts of the only guachimontón-type circular or polygonal compounds, which characterize the nuclear zone of the Teuchitlán tradition in west-central Jalisco where pyrite-encrusted artifacts occasionally form part of the mortuary offerings in the shaft-and-chamber tombs of Late Formative and Early Classic times (between 100 BC and AD 400; see López Mestas C. 2004: 252–253). They were also reported from contemporaneous burials and deposits in the Altos region of northeastern Jalisco (Bell 1974: 154), a territory that was also embedded into the larger network of communication in western and northcentral Mexico at that time.

The earliest occurrence of pyrite-decorated mirrors or adornments from a securely dated and contextualized complex in Zacatecas also comes from Tepizuasco. The initially occupied residential core area of the site on top of the mesa was essentially abandoned around AD 400 and new habitational and ceremonial precincts were built at the foothill portion and on a low east-west oriented promontory linked to the principal hill by a narrow saddle. At the eastern extreme, next to neatly rectangular patio groups, another guachi-montón was erected in the late fourth century AD; this constituted the new ceremonial center and main elite residence of Tepizuasco during the phase of the same name for approximately 150 years until the mid/late sixth century, when this sector itself was abandoned and partially destroyed by fire.

The perfectly circular guachimontón of Compound G unites 10 rectangular platforms in its ring nearly 40 m in diameter. A general east-west orientation is marked by the two principal mounds, more than 4 m high and covering more than 50 m² each. During the test excavations of the western platform, a tomb resembling a miniature wattle-and-daub house was detected at a depth of almost 2 m. The rectangular burial chamber of 1.1 × 1.4 m originally con-tained a multiple burial of at least three adult individuals in supine extended position. However, half of the box-shaped tomb was destroyed during several looting events, leaving only the westernmost portion undisturbed and suffi-ciently intact to recover part of the grave offerings under controlled conditions.

Among the deposited items in this richly endowed funerary space there were four slab-like fragments of dark bluish-gray slate covered with a powdery yellowish substance identified as limonite, which results from the oxidation of iron ores (Schumann 1993: 106–107). This powder represents the remains of the original pyrite coating (figure 8.2). Furthermore, a limonitic sphere (originally a pyrite crystal) still in its rock matrix had been added to the deposit (figure 8.3). All four slate pieces, measuring between 1.5 and 3.5 cm in length, ranged in thickness from 3.5 to 5.5 mm and were fragments of segments cut off from larger disks whose diameters oscillated between 5 and 8 cm. These segments had been sawn off from disks in a deliberate and controlled way, with both the curved and rectilinear edge beveled thereafter by careful grinding. The surface on both sides was only moderately smoothed, probably in order to provide a better grip for the unidentified substance that was used as an adhesive. All segments and fragments had a centrally located perforation drilled from just one side, which would have allowed their use as pendants or beads or even as adornments sewn to a piece of textile or leather.

Given the semi-looted state of the tomb and the fragmentary condition of the four specimens, it cannot be taken for granted that the pyrite-covered slate disks were part of the personal adornments with which the three or more individuals were interred. It is also possible that they were inserted as tes-serae into a larger composite artifact of a ritual nature whose other compo-nents had disintegrated entirely. In spite of these ambiguities, it is most likely that the four mirrors or fragments were in fact part of the elaborate attire of the principal individual and/or his retainers. That the social rank of these

FIGURE 8.2. *Fragments of slate mirror backs, Tomb 4, Tepizuasco, Zacatecas; length of largest specimen: 3.5 cm (photograph by Achim Lelgemann).*

personages was of the highest order is attested not only by the prominent location of their interment, but also by the variability and sophistication of the objects that accompanied the dead: the offering included polychrome and bichrome pottery vessels; small solid figurines, some of them painted, others of a mold-made, disarticulated Teotihuacan type; a unique collection of flat clay seals; items of personal adornment like ceramic earplugs (several of them mold-made, showing human heads in classic Teotihuacan style); beads and pendants made of greenstone, clay, shell, and obsidian; as well as musical instruments and unidentifiable organic specimens. Although no projectile points were recovered, a polished dark stone with a slight groove classified as a dart thrower weight, remnants of atlatl finger loops made from shell, and even fragments of wooden dart or arrow shafts were all present in the assemblage.

The highly specific composition of the mortuary offerings evidently still conforms to the canon established earlier in the shaft-tomb tradition of West Mexico, as do patterns of body treatment and orientation. Definitely not a shaft tomb, Burial 4 of Tepizuasco must be seen as an immediate successor of that funerary complex, an insight which is further confirmed by its relative and absolute chronological position. A series of 17 radiocarbon dates as well as the seriation and cross-dating of diagnostic artifact groups like pottery and figurines leave no doubt that this interment, probably representing

FIGURE 8.3. *Limonitic sphere in rock matrix, Tomb 4, Tepizuasco, Zacatecas; diameter of sphere: 2.5 cm (photograph by Achim Lelgemann).*

one sole funerary event, occurred at the beginning of the sixth century AD, perfectly situated in the Middle Classic Horizon and essentially coeval with Teotihuacan's Late Xolapan or even initial Metepec phase.

In our specific case, it is clear that the four pyrite mirrors mounted on slate slabs should be interpreted at least as indicators of high status, be it social, political, or religious. Although it is true that pyrite (FeS_2) in itself barely constitutes a rare and intrinsically valuable material, the physical properties of the mineral, above all its golden luster and surface polish (Schumann 1993: 154–155), as well as the refined technique invested in their manufacture, make pyrite mirrors or mosaic disks extraordinary markers of individual distinction. As was the case for the earlier specularite specimens found in association with architectural groups on the mesa top, no raw or processed items of pyrite or hematite have been recovered from later commoner residential quarters at Tepizuasco.

Accepting their qualities as indicators of elevated rank, the much more interesting question would be if circular, medallion or even plate sized mosaics, respectively mirrors, might have signaled an even more specific status among the ruling elites. The best-known parallel occurrences in Early and Middle Classic Mesoamerica are definitely the specimens represented in Teotihuacan art (Taube 1992). In addition, physical specimens have been recovered in diverse contexts, most notably the mirror disks attached to the backs of sacrificed warriors buried during the construction of the Quetzalcoatl Temple inside the Citadel (Sugiyama 1989: 90–97). The close analogies found in tombs of Esperanza-phase Kaminaljuyu in the highlands of Guatemala (Kidder et al. 1946: 52–54, 126–134, 155) seem to have the same functions as part of typical warrior accoutrements of the time. Besides, it is generally accepted that these items are the direct forerunners of the *tezcacuitlapilli* of Toltec and Aztec warriors or belligerent deities in Postclassic Nahua-dominated Mesoamerica. During the heyday of Teotihuacan as the undisputed political, economic, and probably military superpower between AD 300 and 500, it would seem that wearing iron-ore-coated plates (as one piece or multicomponent mosaics) reflected at the same time political power, military prowess, and some sort of connection to the prestigious metropolis in the Valley of Mexico, be it as a recognized ally or just simply following an ideal established by the imperial elite in the core region.

It is certainly no coincidence that other Classic-period settlements in West Mexico that have produced (mirror) disks of slate, like Tinganio (or Tingambato) in northwestern highland Michoacán (Piña Chan and Oi 1982: 27, Figure 23) or several sites in the Cuitzeo basin (Filini 2004: 58–60, Figure 5.3u; Macías Goytia 1990: 100), also present distinctive Teotihuacan traits, as for example *talud-tablero* decorated facades or triangular funerary stone masks. At Tepizuasco, the awareness of Teotihuacan's existence and the zeal to at least copy metropolitan lifestyle traits are manifest in the mold-made figurines and ear ornaments (already mentioned as part of the grave offerings in Tomb 4); a freshly struck, unused prismatic blade of green Pachuca obsidian in the same assemblage; and architectural characteristics (a balustrade stairway and an *impluvium* with rectangular pillars at its corners) of an elite residence directly abutting the southern rim of the guachimontón.

LA QUEMADA

The largest quantity known so far of iron-ore tesserae used in the confection of a mirror/mosaic comes from the biggest site in northwestern Mesoamerica, La Quemada, situated in the central section of the Malpaso valley, which forms the upper, northernmost portion of the Juchipila drainage. A lot of 322 pieces of specular hematite was recovered during excavations the author conducted at the Citadel, a Late Classic or La Quemada phase (AD 700–850) patio compound occupying the northern and highest summit of a low mountain that rises to an elevation of 150 m above the surrounding alluvial flood plain of the Malpaso River (Lelgemann 1999). The specimens were obtained through careful screening of the secondary matrix surrounding large and irregularly shaped fill stones inside the pyramid erected at the northeast side of the central courtyard. This procedure became necessary after uncovering a primary extended burial at a depth of slightly less than a meter inside the truncated pyramidal structure. The badly preserved skeleton belonged to a young adult male individual who had been placed directly on some large stone slabs of the fill and originally covered with wooden poles or logs. Five vessels had been deposited on top of them: a red-on-brown painted olla with appliqué nubbins on its shoulder exactly in the center point of the structure and four ring-base bowls in each of the cardinal directions to complete a quincunx pattern. The northern and western bowls were of the pseudo-cloisonné type whereas the southern and eastern goblets were decorated in resist polychrome paint. Scattered all around the vessel deposit were fragments of four more pseudo-cloisonné bowls, all lacking the lower part corresponding to the ring base. In addition to these elaborate ceramic pieces, the assemblage included wooden finger-rings as well as pendants, a quartz crystal, greenstone pendants, shell beads, and pseudo-cloisonné decorated fragments of bottle-gourd containers.

The fine-mesh screening also produced more than 150 pieces of turquoise, both unworked pieces and finished tesserae, as well as five microflakes of translucent green obsidian, one piece significantly larger than the remaining four. There are good reasons to assume that the polished and carefully shaped turquoise tesserae, the specularite chips, and the obsidian flakes had been set originally into one sole multicomponent mosaic on a base disk of perishable material, now totally decayed (figure 8.4). The portion of hematite would have fulfilled the function of a mirror. The vast majority of the 322 specimens obtained were relatively small plaques of 2–3 mm in length and about 0.2 mm thin. The largest piece measured seven mm. All were of irregular shape and did not present beveled edges, contrary to most turquoise tesserae that had been ground into predetermined geometrical forms with slanted rims. On many

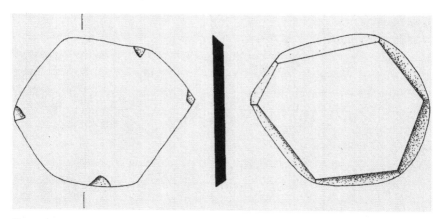

FIGURE 8.4. *Hypothetical reconstruction of specularite-turquoise-obsidian mirror mosaic; Pyramid, Citadel of La Quemada, Zacatecas; diameter: 7 cm (drawing by Achim Lelgemann).*

turquoise and specularite plaques there were still scanty residues of a dark brown organic substance adhering to one of the polished surfaces, probably vegetable juice or resin used as glue for the mosaic surface (Lelgemann 1999: 225–226, photo 58).

Yet another plaque of specular hematite was found in the fill of the inner walkway of the interior loop-shaped wall that surrounds the Citadel on three sides, starting and ending at a bastion-like structure in the middle of the mountain saddle. This isolated piece, certainly a lost specimen that got into the cobble fill as trash, is significantly larger than the minute tesserae of the pyramid fill. It has the form of an irregular hexagon measuring 2 by 1.7 cm with a thickness of 2.5 mm (figure 8.5). The visible surface was finely polished, showing a beautiful dark brown patina with a mottled appearance. Tiny notches at the rim are equidistantly placed in a cardinal pattern that suggests a deliberate modification alluding to the four principal directions. The reverse side was only roughly smoothed, certainly to improve the adherence of the glue. As with most of the turquoise tesserae of the pyramid and many iron-ore plaques of comparable size, all six edges had been beveled. No traces of adhesive could be observed though (Lelgemann 1999: 227, Figure 70a).

It is more than likely that this specimen had formed part of a larger mosaic mirror. Very similar items of pyrite or hematite have been reported from Queretaro (Ekholm 1945) and also from the Chalchihuites area in northwestern Zacatecas (Nárez and Rojas Martínez 1996: 231), the latter being a

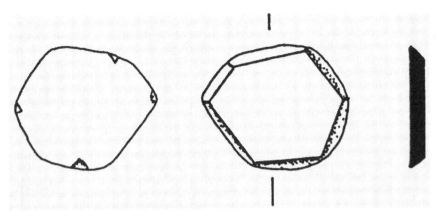

Figure 8.5. *Plaque of hematite; inner wall, Citadel, La Quemada, Zacatecas; maximum length: 2.0 cm.*

complete object with tesselated polygonal plaques mounted on a slate back. Another mirror with these characteristics comes from San Gregorio in the northern highlands in Michoacan (Nárez and Rojas Martínez 1996: 153). In the same state, but further to the south in the heartland of Michoacan's Tierra Caliente province, Isabel Kelly (1947: 125–127) encountered several slate-backed iron-ore mirrors in different contexts. One particular piece is a polygonal plaque of pyrite whose size, technique, and overall form are very similar to the La Quemada specimen (Kelly 1947: 127, Figure 73D). It was part of the grave offerings in Burials 16 and 17 of the San Vicente site in the vicinity of Apatzingán. Kelly assigned the burials to her Tepetate phase, which represents the local Early Postclassic horizon.

After the overall abandonment of major sites and whole regions in west-central Zacatecas toward the end of the Classic period and throughout the Early Postclassic, only the southern canyon lands of the state remained occupied by Mesoamerican farmers, who experienced a far reaching simplification on all levels of cultural life. This reduction and ruralization, accompanied by the disappearance of noble elites and specialized craftspeople working for their demands, might well explain the total lack of sophisticated artifacts in the archaeological record after AD 1000, including iron-ore mirrors and mosaics that were no longer manufactured. A possible exception might have been the southernmost sector of the Tlaltenango valley and its foremost ceremonial center, El Teul, whose inhabitants managed to establish close ties with Aztatlan complex populations along West Mexico's Pacific coastline and some sites in the central lake district of Jalisco. The continued production of high-quality craft items like polychrome pottery has been confirmed recently by ongoing INAH excavations at the site and may well turn up artifacts of pyrite and specularite, items that are present at neighboring areas in Postclassic contexts.

The most notable find because of its direct association with turquoise in a ceremonial deposit was the discovery of a disintegrated mosaic (mirror) in a cache underneath the only intact Chac Mool of Tula in Hall 2 of Structure 3 ("Palacio Quemado") during the large-scale exploration under Jorge Acosta (1964: 53, 56, plates XI-XIII) of the 1950s and 1960s. More recently, a comparable deposit containing pyrite in association with the famous warrior tunic was detected by Cobean in the same building when the original stucco floors of Hall 2 were broken for deep-test trenching of the platform fill (Cobean and Estrada Hernández 1994).

The only specimens of iron ore (pyrite) known for the northern extremes of Mesoamerica of the Postclassic era remain the objects deposited in a Period VI burial excavated by Ekholm at Las Flores, Tampico, in the Huasteca region (Ekholm 1944: 489, Figure 51e). They must have been imported, as the sedimentary soils of the Pánuco River floodplain do not contain noteworthy mineral crystals of any sort.

DISCUSSION

As items of personal embellishment and social distinction, the undisputed significance of iron ore mirrors and mosaics in pre-Columbian cultures of Mesoamerica and other cultural provinces of the continent (Ekholm 1973) is confirmed by the Zacatecas material presented here. The very specific contexts in which specularite- and pyrite-coated disks were found at major archaeological sites within the state allow further inferences, as they imply multiple dimensions of function and meaning. Especially the occurrence of a multi-component exemplar (with specularite, turquoise, and obsidian) in a ceremonial deposit in the pyramid of the Citadel at La Quemada needs further contemplation due to the extraordinary features of the structure, the compound it belongs to, and the burial it contained.

As several tesserae of the composite mosaic-mirror were screened from the matrix above the olla and goblets set, the object cannot have been part of the personal adornment or accoutrement of the individual buried underneath the vessels. The item constituted rather the topmost of the remaining offerings and might be seen as a sort of lid or cover of the whole assemblage. Even as such, it was an integral part of the deposit that was laid down when the building process concluded and is therefore to be interpreted as a dedicatory cache. Although it cannot be ruled out completely that at a later date—while the structure was in use—a high-ranking personage was given a pompous burial at a prominent location that included a set of valuable and sumptuous offerings, the alternative appears less likely considering that no disturbance of the fill matrix could be observed.

Among the dedicatory items, a quite unusual trait of the body treatment must be correlated directly to the presence of at least one mirror, possibly even two or more (for a detailed discussion of the burial see Lelgemann 1999: chapter 10): the lower right leg of the young adult male had been either cut off or ripped away from the upper part and laid crosswise above the thigh. The shafts of the femur and the tibia/fibula crossed almost exactly (at a distance of 7 cm) at the pyramid's center point. That this was no coincidence was corroborated by the equally central location of the olla directly above the crossed bones and the deposition of the four goblets at the extremes or epiphyses of the three shafts. Thus both the legbones and the vessel deposit clearly marked a quincunx or directional association. Under these circumstances, it is to be remembered that the five obsidian chips probably inserted in the mirror created the same cosmogram. The association of a young man with an amputated leg and a hematite-turquoise-obsidian mirror in a Northwest Mexican monumental site inevitably leads to the assumption that the ritual offering and the sanctuary where it was placed were once dedicated to a deity; probably the immediate forerunner or prototype of the god Tezcatlipoca, venerated by most Nahua-speaking groups in Central Mexico during the Postclassic period. The severed leg and the mirror(s) were the defining attributes of that divinity, who was also associated with the night (sky), the North, and winter (see Olivier 1997 for a complete study of the Mesoamerican Tezcatlipoca complex). The Citadel of La Quemada is located at the northernmost ceremonial precinct of the site and has a precise orientation to one of the two solstice axes (Lelgemann 1997). The frontal staircase of the pyramid directly faces the sunset at the winter solstice on December 21, that is, the shortest day and respectively the longest night of the solar year. Both facts add strong supportive evidence to the interpretation of the pyramid deposit as a cache related to Tezcatlipoca or rather his Late Classic Northwest Mesoamerican antecedent.

This argument gains further weight through correlating the offering of La Quemada to the coeval burial in the northern corner of the Hall of Columns at Alta Vista, which since its discovery has been linked to the Tezcatlipoca cult by the excavators (Holien and Pickering 1978). Further mirrors of diverse material and association have been found in mortuary contexts (Kelley and Kelley 1980) and dedicatory offerings at Alta Vista (Medina González and García Uranga 2010). Obsidian mirrors in Postclassic Central Mexico, normally carved from one piece and showing one or two highly polished surfaces, are often considered the peculiar devices of vision and reflection that are an important component of the Tezcatlipoca complex. At La Quemada one item of this sort had already been found by the pioneer of Zacatecas archaeology, Carl de Berghes (1996 [1855]: 27, plate IX, Figure 2), during his 1830/31 explorations of the site's western flank: the perfectly round obsidian mirror—which has since been lost—had an eye for suspension and seems to have had approximately the same size as the specimen from the Citadel pyramid. Obsidian is generally scarce in the archaeological record of La Quemada, although the site was fully integrated into supraregional exchange networks (Darling and Glascock 1998) and received volcanic glass from multiple sources in Central, Western, and Northern Mexico (Trombold et al. 1993). The obsidian chips set into the citadel mirror/mosaic were of the translucent green variety of the Cerro de las Navajas flow near Pachuca, Hidalgo, which was the southeasternmost source of obsidian channeled to La Quemada. In this respect, it seems to be meaningful that the other principal component in the mosaic was turquoise, another luxury raw resource imported through well-organized trade relations from deposits situated in Chihuahua and New Mexico (Weigand et al. 1977; Weigand and Harbottle 1992), that is, the northwestern extremes of the exchange system. We will never know for sure whether the makers of the mirror were aware of the exact geographical proveniences of the raw material they used in the manufacture and if they combined them on purpose in one object that was certainly in itself a miniature cosmogram, as was the pyramidal substructure and in fact the patio compound as a whole. Provided that they chose these materials deliberately to represent the opposite ends of their physical and conceptual universe, the third component, specularite of local origin, would have symbolized the central location.

Beyond these horizontal-geographical or rather cosmological implications, another dimension might be added when considering possible significances suggested by the distinct colors of the combined minerals: turquoise stands for the deep blue of the diurnal, cloudless skies, and therefore by ideational association reflects the upper celestial sphere, the sun at zenith, the male, aggressive principle, and in Nahua ideology a strong undertone of war and warrior glorification. On the contrary, obsidian is decidedly related to the exact opposites: night and darkness, cold and humid, the moon or even the dead sun in the underworld, the female counterpart, and a passive, enduring posture.

Within this vertical conceptualization of realms and correlated principles, iron ores (hematite and pyrites in particular) would represent the middle level, the chthonic surface which matches perfectly the natural occurrence of these minerals (for Mesoamerican concepts of space and their associations see Carrasco 1990; López Austin 1996).

Various additional functional or symbolic facets of the Zacatecas specimens, particularly the find at the La Quemada Citadel, have been taken into consideration but were ultimately discarded for several reasons. The potential use of concave mirrors as devices to kindle fires seems to be out of the question for all known specimens of Zacatecas, which were evidently flat. Even if the mirror mosaic from the citadel pyramid had been mounted on a concave base (e.g., the interior of a gourd) the more than three hundred individual pieces put together would have resulted in a quite uneven surface and the consequently multiple refraction of sunbeams would have prevented their necessary bundling in a single focal point. The same physical phenomenon of multidirectional refraction would not have produced a clear image of any object to be contemplated on the reflective surface. Besides, the generally reduced size of the artifacts known so far cannot have been practical for that purpose. Finally, a very possible integration of mirrors as devices for creating visions and transformations in the course of shamanistic practices is too complex a topic to be discussed further within the limited space of this chapter (see chapter 9 in the present volume).

As a concluding comment, it must be stated that from a technical and aesthetic point of view, the Zacatecas specimens discussed above cannot compare to the finest iron-ore mirrors and mosaics known from the nuclear regions of Mesoamerica, where they often bear additional decoration, mostly on their reverse sides, either painted or carved (von Winning 1990). Nonetheless, the Zacatecas iron-ore mosaics do appear in a wide range of different compositions and settings where they proved to be key diagnostic elements identifying social status, craft specialization, interregional trade relations, religious cult complexes, and even specific deities of the Mesoamerican pantheon. It is hoped that this chapter and volume will help future researchers to pay closer attention to these undeservedly undervalued items, often (de)classified as "minor" artifacts or simply not retrieved, analyzed, or published at all.

REFERENCES

Acosta, Jorge R. 1964. "La decimotercera temporada de exploraciones en Tula, Hgo." *Anales del Instituto Nacional de Antropologia e Historia* 16: 45–76.

Bell, Betty. 1974. "Excavations at El Cerro Encantado, Jalisco." In *The Archaeology of West Mexico*, ed. Betty Bell, 147–167. Ajijic, Mexico: West Mexican Society for Advanced Study.

Berghes, Carl de. (Original work published 1855) 1996. *Descripción de las ruinas de asentamientos aztecas durante su migración al Valle de México, a través del actual Estado Libre de Zacatecas.* Traducción, estudio introductorio y notas de Achim Lelgemann, Joyas Bibliográficas Zacatecanas II, Facultad de Humanidades, Universidad Autónoma de Zacatecas, Centro Bancario del Estado de Zacatecas. Zacatecas: Gobierno del Estado de Zacatecas.

Carrasco, David. 1990. *Religions of Mesoamerica: Cosmovision and Ceremonial Centers.* San Francisco: Harper and Row.

Cobean, Robert H., and Elba Estrada Hernández. 1994. "Ofrendas toltecas en el Palacio Quemado de Tula." *Arqueología Mexicana* 6: 77–78.

Darling, J. Andrew, and Michael D. Glascock. 1998. "Acquisition and Distribution of Obsidian in the North-Central Frontier of Mesoamerica." In *Rutas de Intercambio en Mesoamérica, III Coloquio Pedro Bosch Gimpera,* edited by Evelyn Childs Rattray, 345–64. México: Universidad Nacional Autónoma de México.

Ekholm, Gordon F. 1944. *Excavations at Tampico and Panuco in the Huasteca, Mexico.* Anthropological Papers of the American Museum of Natural History, Vol. 38, Part 5. New York: American Museum of Natural History.

Ekholm, Gordon F. 1945. "A Pyrite Mirror from Querétaro, Mexico." *Notes on Middle American Archaeology and Ethnology* 53: 178–181.

Ekholm, Gordon F. 1973. "The Archaeological Significance of Mirrors in the New World." *Proceedings of the 40th International Congress of Americanists* 1: 133–136.

Filini, Agapi. 2004. *The Presence of Teotihuacan in the Cuitzeo Basin, Michoacán, Mexico: A World-System Perspective.* BAR International Series 1279. Oxford: Archaeopress.

Macías Goytia, Angelina. 1990. *Huandacareo: Lugar de juicios, tribunal.* Colección Científica 222. México City: INAH.

Holien, Thomas, and Robert B. Pickering. 1978. "Analogues in Classic Period Chalchihuites Culture to Late Mesoamerican Ceremonialism." In *Middle Classic Mesoamerica, A.D. 400–700,* ed. Esther Pasztory, 145–157. New York: Columbia University Press.

Kelley, Ellen Abbott, and J. Charles Kelley. 1980. "Sipapu and Pyramid Too: The Temple of the Crypt at Alta Vista, Chalchihuites." *Transactions of the Illinois State Academy of Science. Illinois State Academy of Science* 73: 62–79.

Kelly, Isabel T. 1947. *Excavations at Apatzingán, Michoacán.* Viking Fund Publications in Anthropology 7. New York: Wenner-Gren Foundation for Anthropological Research.

Kidder, Alfred V., Jesse D. Jennings, and Edwin M. Shook. 1946. *Excavations at Kaminaljuyu, Guatemala.* Publication no. 561. Washington, DC: Carnegie Institution.

Lelgemann, Achim. 1997. "Orientaciones astronómicas y el sistema de medida en La Quemada, Zacatecas, México." *Indiana* 14: 99–125.

Lelgemann, Achim. 1999. *Die Zitadelle von La Quemada, Zacatecas: Archäologische Untersuchungen in einem spätklassischen Patio-Komplex der nordwestlichen Peripherie Mesoamerikas.* Berlin: Freie Universität Berlin.

Lelgemann, Achim. 2009. "Sinopsis de investigaciones arqueológicas recientes en el Noroeste de Mesoamérica." *Espaciotiempo* 3: 123–140.

Lelgemann, Achim. 2010. "El Formativo Terminal y el Clásico Temprano en el Valle de Malpaso-Juchipila (Sur de Zacatecas)." In *El sistema fluvial Lerma-Santiago durante el Formativo y el Clásico Temprano: Precisiones cronológicas y dinámicas culturales; Memoria del segundo seminario-taller sobre problemáticas regionales*, edited by Laura Solar Valverde, 181–205. Colección Científica 565. México: INAH.

López Austin, Alfredo. 1996. "La cosmovisión mesoamericana." In *Temas mesoamericanos*, edited by Sonia Lombardo and Enrique Nalda, 471–507, México: INAH.

López Mestas C., Lorenza. 2004. "Costumbres funerarias en el centro de Jalisco." In *Tradiciones arqueológicas*, edited by Efraín Cárdenas García, 243–259. El Colegio de Michoacán, Zamora, Morelia: Gobierno del Estado de Michoacán.

Medina González, José Humberto, and Baudelina L. García Uranga. 2010. *Alta Vista. A cien años de su descubrimiento.* Gobierno del Estado de Zacatecas, Consejo Nacional para la Cultura y las Artes. México: INAH.

Monografía geológico-minera del estado de Zacatecas. 1991. Consejo de Recursos Minerales, Subsecretaría de Minas e Industria Básica. México: Secretaría de Energía, Minas e Industria Paraestatal.

Nárez, Jesús, and José Luis Rojas Martínez. 1996. *Sala de las Culturas del Norte de México.* Colección Catálogos, Museo Nacional de Antropología. México: INAH.

Olivier, Guilhem. 1997. *Moqueries et métamorphoses d'un dieu aztèque: Tezcatlipoca, le "Seigneur au miroir fumant."* Mémoires de l'Institut d'Ethnologie 33, Musée de l'Homme. Paris: Muséum Nationale d'Histoire Naturelle.

Piña Chan, Román, and Kuniakí Oi. 1982. *Exploraciones arqueológicas en Tingambato, Michoacán.* México: INAH.

Schiavitti, Vincent W. 1996. *Organization of the Prehispanic Suchil Mining District of Chalchihuites, Mexico, A.D. 400–950.* Buffalo: State University of New York.

Schumann, Walter. 1993. *Handbook of Rocks, Minerals, and Gemstones.* New York: Houghton Mifflin.

Síntesis Geográfica de Zacatecas. 1981. México: Instituto Nacional de Geografía, Estadística e Informática.

Sugiyama, Saburo. 1989. "Burials Dedicated to the Old Temple of Quetzalcoatl at Teotihuacan, Mexico." *American Antiquity* 54 (1): 85–106. http://dx.doi.org/10.2307 /281333.

Taube, Karl A. 1992. "The Iconography of Mirrors at Teotihuacan." In *Art, Ideology, and the City of Teotihuacan*, ed. Janet C. Berlo, 169–204. Washington, DC: Dumbarton Oaks Research Library and Collection.

Trombold, Charles D., James F. Luhr, Toshiaki Hasenaka, and Michael D. Glascock. 1993. "Chemical Characteristics of Obsidians from Archaeological Sites in Western Mexico and the Tequila Source Area: Implications for Regional and Pan-Regional Interaction within the Northern Mesoamerican Periphery." *Ancient Mesoamerica* 4 (2): 255–270. http://dx.doi.org/10.1017/S0956536100000948.

von Winning, Hasso. 1990. "Altmexikanische Pyritspiegel mit reliefierter Rückseite." In *Circumpacifica: Festschrift für Thomas S. Barthel*, ed. Bruno Illius and Matthias Laubscher, 455–481. Bern: Peter Lang.

Weigand, Phil C. 1968. "The Mines and Mining Techniques of the Chalchihuites Culture." *American Antiquity* 33 (1): 45–61. http://dx.doi.org/10.2307/277772.

Weigand, Phil C. 1982. "Mining and Mineral Trade in Prehispanic Zacatecas." In *Mining and Mining Techniques in Ancient Mesoamerica*, edited by Phil C. Weigand and Gretchen Gwynne. Special Issue Anthropology 6: 87–134.

Weigand, Phil C., Acelia García de Weigand, and J. Andrew Darling. 1999. "El sitio arqueológico 'Cerro de Tepecuazco' (Jalpa, Zacatecas), y sus relaciones con la tradición de Teuchitlán (Jalisco)." In *Los Altos de Jalisco a fin de siglo, Tercer Simposium, Memorias*, ed. Cándido González Pérez, 241–274. Guadalajara: Universidad de Guadalajara.

Weigand, Phil C., and Garman Harbottle. 1992. "The Role of Turquoise in the Ancient Mesoamerican Trade Structure." In *The American Southwest and Mesoamerica. Systems of Prehistoric Exchange*, ed. Thomas G. Baugh and John E. Ericson, 159–177. New York: Plenum Press.

Weigand, Phil C., Garman Harbottle, and Edward V. Sayre. 1977. "Turquoise Sources and Source Analysis: Mesoamerica and the Southwestern U.S.A." In *Exchange Systems in Prehistory*, ed. Timothy M. Earle and John E. Ericson, 15–34. New York: Academic Press. http://dx.doi.org/10.1016/B978-0-12-227650-7.50008-0.

9

The ancient Maya created one of the most elaborate civilizations of pre-Columbian America. Their complex, hierarchical society evolved and flourished in the tropical lowland rainforests of Central America during the Classic period ca. AD 250–900. They developed a rich cosmology and formal religious system, which can be partially reconstructed today using methods of archaeology, art history, and ethnohistory. This chapter considers complex lithic artifacts termed "mirrors," elucidating their sociopolitical and ideational functions for the ancient Maya individuals who possessed and used them. Moreover, I advance a hypothesis and evidence suggesting that the ancient Maya harnessed certain psychoactive substances to augment the effectiveness of their mirrors for consulting (what they believed to be) a spiritual Otherworld.

Techniques of Luminosity

Iron-Ore Mirrors and Entheogenic Shamanism among the Ancient Maya

MARC G. BLAINEY

ANCIENT MAYA IRON-ORE MIRRORS AS SHAMANISTIC TOOLS

Despite the prevalence of iron-ore mirror remains, which are found throughout ancient Mesoamerica, Karl Taube's analysis of mirrors at the site of Teotihuacan in Central Mexico highlighted a deficiency of analysis concerning these artifacts in the Maya subarea (Taube 1992). Described by Richard Woodbury (1965: 172) as "masterpieces of the stoneworker's craft," the ancient Maya mirror type consists of a mosaic surface made of polished iron-ore fragments fitted and adhered to a

DOI: 10.5876/9781607324089.c009

solid backing of wood, ceramic or, most commonly stone. My MA thesis, published online with the European Mayanists Association (Blainey 2007), sought to answer Taube's appeal by exploring the political, ideological, and economic functions of mirrors within ancient Maya life. I assessed the archaeological provenance of over 500 mirror artifacts as well as over 50 instances of mirrors depicted on painted ceramic vessels. All of these specimens are compiled together into a detailed database (Blainey 2007: 205–224). A condensed version of this thesis was published recently, coauthored with Paul Healy (Healy and Blainey 2011). One also witnesses the upsurge of archaeological interest in ancient Mesoamerican mirrors in articles by Nielsen (2006) and Nelson et al. (2009), as well as in the Wikipedia article on the topic.

In reviewing the data acquired from site reports over the last century of Mayanist archaeology, I documented mirrors dating from the Middle Preclassic period to the Contact period and found in all subdivisions of the Maya subarea except the Pacific Coastal Plain and Piedmont regions (see Blainey 2007). These mirrors occur most often in elite burial and cache contexts, usually within the monumental architecture of site cores. Although there are a small number of mirrors (mostly fragments) recovered from structural fill and surface contexts, their regular association with sumptuous interments along with jade, decorated pottery, and jewelry of bone, shell, and stone, situates these objects overwhelmingly with elite provenances. This redundancy in the archaeological contexts of iron-ore mirrors insinuates that they were "prestige items" controlled by the noble classes, most likely owned by the same elite individuals with which they were interred. According to Colin Renfrew (1986: 161), the ownership of such an item, "by virtue of the prestige it confers . . . offers access to social networks and to other resources that are closed to those lacking such prestige." Ancient Maya elites' possession and public display of prestige items symbolized their divine right to power.

A consistent pattern emerged through the course of my research into the archaeological and iconographic records of the ancient Maya, denoting supernatural uses of iron-ore mirrors by members of the royal court. Probably related to a more rudimentary practice of "water scrying" with bowls of liquid, the use of mirrors in "divinatory scrying" indicates that some ancient Maya elites served in the capacity of shaman (see Blainey 2007: 17–34; Blainey 2010; Healy and Blainey 2011). The shaman-ism notion is derived from the word šaman, meaning "one who is excited, moved, raised" in the language of the Tungus (Evenki) peoples of Siberia, and entered into Western parlance during the eighteenth century through the writings of European and Russian explorers (Eliade 1964: 495–496; Walsh 2007: 13; Znamenski 2007:

viii, 3–5). The concept of shamanism has since gained popularity as an expedient device for interpreting and classifying a variety of spiritual practices (Narby and Huxley 2001). Persisting as a contentious issue among anthropologists and archaeologists, the religious category of shamanism is generally defined by techniques and ritual behavior that are conducted with the intention of inducing altered states of consciousness (Freidel et al. 1993; Hutson 1999; Kehoe 2000; Klein et al. 2002; Pearson 2002; Price 2001; Prufer 2005; Winkelman 2000). When it comes to interpreting ancient Maya mirrors, I prefer the cross-culturally inclusive strategy evoked in Joan Townsend's (2001) definition of the role of the shaman: "A shaman is one who has direct communication with spirits, is in control of spirits and consciousness, undertakes some (magical) flights to the spirit world, and has a this-material-world focus rather than a goal of personal enlightenment." The use of mirrors is closely tied to shamanistic practices documented ethnographically in Asia. The extant literature makes repeated reference to northeast Asian shamans' strong associations with mirrors, noting that historically "the shamans' omniscience stemmed from the power of their mirrors" (Litvinskii 1987: 559). It is common for metal mirrors to be included as a component of specialized costumes worn by shamans in traditional East Asian cultures. Eliade describes how "the mirror is said to help the shaman to 'see the world' (i.e., to concentrate), or to 'place the spirits,' or to reflect the needs of mankind, and . . . looking into the mirror, the shaman is able to see the dead person's soul" (Eliade 1964: 154; quoted in Saunders 1988: 20). Scholars of Siberian religious systems also note that mirrors were used across this region to observe or communicate with spirits and to foretell the future (Walter and Fridman 2004: 179, 582, 606, 793). These ethnographic observations are in agreement with the archaeological and iconographic evidence examined below, indicating that some elements of Siberian shamanism accompanied pre-Columbian migrations across the Bering land bridge. The present text ventures to refine our understanding of iron-ore mirrors as corroboration of shamanistic practices in ancient Maya society.

THE *REFLECTIVE SURFACE COMPLEX*: ICONOGRAPHIC AND LINGUISTIC CLUES

In his article "Anthropological Reflections on Archaeological Mirrors" Nicholas Saunders (1988: 4) refers to shamanistic beliefs in positing that "the translation of a metaphysical belief into a physical object links the mirror and its material form with a wider concept surrounding notions of reflected images and parallel spirit worlds." He infers obvious associations that mirrors have with the surfaces of water and eyes, particularly the eyes of jaguars (Saunders 1988: 9–12). These associations with eyes, the surface of water, and the mirrors themselves constitute what I have termed a *reflective surface complex* of ancient Maya worldview, whereby all luminescent surfaces are perceived as exceptional renderings of the liminal threshold between the natural and spiritual realms (Blainey 2007; Healy and Blainey 2011: 234–238). The ancient Maya reflective surface complex is closely associated as a subcategory of the larger divinatory casting and scrying complex identified by Taube (1983: 120–121) for Teotihuacan. Indeed, at Teotihuacan, the casting of seeds and maize kernels is commonly associated with mirrors in their visual iconography. Saunders (1988: 13–22) sees a connection between the materialization of these liminal portals and the restricted control of such objects as a universal trait of shamanism. He suggests that as shamanistic tools, mirrors help to "reinforce ideological activity," and that they maintained their utility for emerging elites in the development of social complexity, as can be seen in the archaeological records of the Olmec and Maya civilizations.

Numerous images on ancient Maya painted polychrome ceramics illustrate iron-ore mirrors functioning as principal objects in the royal court, an observation that matches the archaeological prevalence of mirror artifacts occurring in high-status burials and special/ritual caches in elite structures. As for their more precise functions within this elite context, the iconographic evidence overwhelmingly suggests that the mirrors were at times meant to be gazed into, but what this gazing indicates is a much more elusive consideration. A major clue as to the emic function of mirrors comes from scenes depicting what are probably supernatural beings partaking in ritual behaviors that are, for the most part, esoteric and beyond the reach of modern understanding. For example, on vessel K559 in the Justin Kerr (2013) database the Moon Goddess is giving birth to a rabbit, a symbol of the moon, onto (or in front of) a mirror situated directly where the cosmic World Tree grows; however, the exact meaning of the mirror in this scene is difficult to decode.

On the other hand, the images on K2929 and K5944 are very revealing in terms of the functions of ancient Maya iron-ore mirrors. It is important to note that ancient Maya iconography utilizes synaesthesia, where the modality of sight gleans signs that are intended to carry meaning, sound, and scent (Houston and Taube 2000: 263). In K2929, the mirror partitions a human figure and an anthropomorphic rabbit-jaguar[1] (figure 9.1). The latter character can be interpreted as a Maya *wahy* spiritual co-essence (Calvin 1997; Grube and Nahm 1994; Houston and Stuart 1989) with which the human figure is

FIGURE 9.1. *Detail of painted vessel K2929, depicting the communication of an anthropomorphic rabbit-jaguar (left) with a human figure (right) via a mirror (propped up and viewed from the side at center) (Kerr 2013).*

communicating via the mirror. Hereby, the mirror appears to act as a bridge between two normally divided worlds. This idea of the mirror as a communication device between the human realm and the spiritual Otherworld is strengthened by the fact that the rabbit-jaguar *wahy* is emitting speech scrolls above and across the mirror toward the human figure. K5944 (not shown) depicts a similar situation, but with a water-bird *wahy* replacing the rabbit-jaguar *wahy*. The water-bird in K5944 is denoted by the stylized T501 *imix* glyph (representing "a water lily blossom" [Montgomery 2002: 97]) composing its wing. It is apparent that the same T501 *imix* symbol is incorporated into the ear of the rabbit-jaguar in K2929 (the possible significance of associating this water lily sign with the mirror and *wahy* characters in these pictorial scenes will be reviewed below).

Another tool of power wielded by Maya elites was their elaborate system of writing. Of particular interest for this discussion is the elliptical glyph encapsulating one, two, or three curved bands, appearing as both a supplemental affix (T24) and as a full logograph (T617). With curved bands either empty or crosshatched, it appears most frequently in its T24 affix role, 🔲, often as part of a logograph whose meaning is not immediately related to mirrors, interpreted more generally as symbolizing a "celt" or "reflective stone" (Macri and Looper 2003). The polyvalent nature of the Mayan script is evident in the apparent use of T24/T617 to indicate the vague notion of "shininess," where its occurrence can also denote, for example, obsidian blades, celts, or even shell tinklers. Furthermore, the T24 phonetic version of the "shininess" glyph stands for the syllable *-il* or sometimes *-li*. When this same sound *il* is a morpheme for the verb "to see" (Houston et al. 2006: 139), it is represented by a logogram showing the cross-section of an eyeball with lines of vision radiating from the inner chamber: or 🪞. The past-tense version of this verb is *ilaj*, 🪞 "was seen," and provides an overt connection with the flared-out mirror-backs seen in painted pottery scenes (figure 9.2). As observed by Caroline Humphrey (2007: 187), we see similar concepts expressed by modern Mongolian shamans, where "the shaman's eyes in trance do not see but become akin to the mirror when it operates as a magical light-flashing weapon of those spirits inside or behind it"; the clear associations between the "shiny" objects and eyes in Maya epigraphy are thus akin to the Mongolian tendency "to liken the mirror to the eyes [because] in both cases they gather in and absorb images . . . from outside and cast them out again from inside." We see this association in a recent reanalysis of the notorious Maya 2012 Bak'tun end-date recorded on Tortuguero Monument 6 (see Blainey 2011). Gronemeyer and MacLeod (2010: 11–15) have employed updated photographic imagery to show that on glyph block P4, which was previously assumed to be effaced, the "mirror glyph" is scarcely discernible. These authors now translate this glyph block as "a seeing," so that the event prophesized to occur on the Winter Solstice in December 2012 can be read as "it will happen, a 'seeing'."

In linking the archaeological and epigraphic evidence, we must consider traits shared across the different dialects of the Mayan language family. Reviewing dictionaries of Conquest-period and modern Mayan dialects, Schele and Miller (1983: 12–14) identified multiple examples where the terms *nen* and *lem* function "as the root for [expressions such as] 'lightning,' 'gleam,' and 'shine.'" They discovered that variants of *nen* and *lem* during the Contact period also refer to the concepts of "rulers and persons of importance as 'the reflection of the world' . . . 'succession in office' . . . 'to think,' 'to imagine,' 'to contemplate,' and 'to meditate.'" This linguistic data is complemented by the occurrence of a phonetic spelling for the same word (ne-na, *nen*) in a text carved onto a mirror-back from Topoxte, Guatemala; here the full phrase is *u-nen*, translated as "his/her mirror" or "his/her shiny surface" (Fialko 2000; see also Grofe 2006). This semantic continuity, from pre-Hispanic period hieroglyphs to Conquest-period linguistics, acts to fortify the current proposition that mirrors (shiny, gleaming artifacts) served as archetypes for a broader cognitive domain encompassing all reflective surfaces found in the natural and artificial worlds. Referring to the T24/T617 glyph as an icon of the reflective surface complex, it may represent a mirror or any other shiny surface that the ancient Maya would have understood as a liminal portal linking the human world with the spiritual Otherworld.

FIGURE 9.2. *Detail of painted vessel K625, showing a propped-up mirror (left) with its flared-out back (Kerr 2013).*

The precise techniques that ancient Maya ritual specialists used to unlock perceptual barriers between the human and spirit realms are not recoverable archaeologically. However, I will now propose a commonsensical scenario explaining how the use of certain psychoactive plants and mushrooms would have potentiated the experiential phenomenology of scrying with iron-ore mirrors. Rather than referring to "drugs," "hallucinogens," or "psychedelics," terms that are culturally loaded with biases from our own modern Western approach, I prefer the term "entheogen" (see McGraw 2004: 204–251). *Entheogen* means "plants or chemical substances which awaken or generate mystical experiences" (Forte 1997: 1). Thus, because the word *entheogen* portrays these substances empathically, recognizing those psychoactive substances used for religious or shamanic purposes (Ruck et al. 1979: 146), it is a more appropriate term when discussing the ancient Maya.

ENTHEOGENIC MODIFICATIONS OF
CONSCIOUSNESS IN ANCIENT MESOAMERICA

While evidence for the ceremonial use of entheogens is often discounted, a handful of scholars have pointed to conspicuous artifacts suggesting that this practice was present among ancient Mesoamericans (see Blainey 2005; Carod-Artal 2011). In the most prominent example of this work, Marlene Dobkin de Rios (1974) outlined how the archaeological record implies that the ancient Maya purposefully ingested psychoactive substances now viewed by modern Western society as dangerous "hallucinogens." As expected, in the replies that follow Dobkin de Rios's article, there are those for and against her (somewhat speculative) hypotheses, but responses from luminaries of Maya archaeology are especially disapproving; for instance, Tatiana Proskouriakoff states that "direct flights of fancy such as [Dobkin de Rios] offers here should have no place in a scholarly journal, and I think that the referees have been guilty of a serious error of judgement" (Dobkin de Rios 1974: 159). Such admonishments demonstrate a sociopolitical bias on the part of many Mayanists, as recognized in Nicholas Hellmuth's reply (Dobkin de Rios 1974: 155–156):

> Dobkin de Rios provides a needed kick to conservative Maya scholars, most of whom still view the ancient Maya as calm, benign people as described by early Carnegie Institution of Washington writers. Sylvanus Morley would be shocked to learn that his Maya may have taken drugs . . . Anthropology's wide-open view of the diversity of mankind's ways should instead allow us to look at some possibly useful insights from Dobkin de Rios's paper.

However, as Dobkin de Rios points out, statistical surveys have shown that researchers tend to overlook the importance of entheogens within a culture until they experiment with the chemicals themselves (Dobkin de Rios 1974: 160–161). Accordingly, the following section seeks to renew earnest scholarly explorations of the psychoactive species potentially exploited by the ancient Maya to achieve shamanistic states of mind.

For instance, the water lily was "among the more popular motifs at Copan . . . [and] was an important element of royal symbolism . . . frequently found as part of the headdress of Maya elites, gods, and figures associated with the Underworld" (Davis-Salazar 2003: 278). As discussed earlier, the *imix* glyph (representing a water lily) is incorporated on the bodies of zoological *wahy* spirits depicted on painted pottery images involving mirrors as supernatural communicatory devices. The iconographic prominence of the water lily spurred assertions that the species *Nymphaea ampla* was prized by the ancient Maya for its psychoactive properties (Emboden 1981, 1982). It has been presumed that "the alkaloids apomorphine, nuciferine, and nornuciferine, isolated from the rhizomes of *N. ampla*, may be responsible for the psychotropic activity" (Schultes et al. 2001: 66–67). However, it should be noted that effects of *Nymphaea* water lily species on human perception are not "hallucinogenic,"

FIGURE 9.3. *Mushroom stones.*

as has been reported (Cano and Hellmuth 2008: 1; FLAAR 2008: 2), but rather smoking or swallowing the flowers of these plants produces more of a "mild sedative effect" (Erowid 2006). Of course, knowledge of the tranquillizing properties of water-flowers dates back thousands of years in Western literature, at least to the daydreaming "lotus eaters" encountered in Homer's *Odyssey* (Griffiths 2010: 32–33).

Whereas use of the water lily is conjectural, it is quite apparent that the ancient Maya were at some point ceremonially consuming entheogenic mushrooms. There are over 20 different species of mushrooms found in Mexico that contain mind-altering chemicals, most often the alkaloid psilocybin (4-phosphoryl-DMT), and Dobkin de Rios (1974: 148, 153) has made "a good case" that some of these species were employed and perhaps even traded by the ancient Maya. Archaeological evidence for the use of these mushrooms is indicated by the numerous mushroom-shaped sculptures found at archaeological sites across Mesoamerica (Blainey 2005; de Borhegyi 1961; de Borhegyi 1963; Wasson 1980; see also de Borhegyi and de Borhegyi-Forrest 2013). These "mushroom stones" (figure 9.3) were made of both stone and ceramic, often including an anthropomorphic or zoomorphic effigy build into the base, forming the stipe (or stalk) of the mushroom; some were recovered in association with stone *manos* and *metates* (grinding tools) from sites in the Guatemalan Highlands and Pacific piedmont (Miller and Taube 1997: 90). Considering that a Oaxacan Mazatec woman was observed in 1960 using a mano and metate to grind psilocybin mushrooms prior to ingesting them in shamanistic rituals (Wasson 1980: 175–197), it is conceivable that similar mushroom rites were performed by the ancient Maya who carved these effigy stones.

Specific archaeological evidence for this practice comes from Mound E-III-3 at Kaminaljuyu, where archaeologists found "three fragments of the heads of 'mushroom' stones" in the mound fill. Moreover, Shook and Kidder (1952: 111–112, Figure 78) found "a very finely made tripod" mushroom stone in Tomb 1 of this mound, a context that also included five pairs of manos and metates. Three of these metates are shaped like a toad or frog. As insinuated by Dobkin de Rios (1974: 148–150; see also Blainey 2005: 11–14), it could be that entheogenic mushrooms were being mixed on manos and metates with the secretions from the parotid glands of some toad species (containing the entheogenic chemical *bufotenine*). Shook and Kidder (1952: 116) surmise that an iron-ore mirror, found in this same mound context with so much explicit entheogenic imagery, was originally placed on the roof of the tomb upon its initial closing.

It has been further suggested that tobacco, alcohol, and/or psychoactive flowers like those of the water lily were administered during ancient Maya enema rites (de Smet 1985; de Smet and Hellmuth 1986). Although the exact identity of ritual enema ingredients remains undetermined, Stross and Kerr (1990) surmise that enemas likely contained alcohol infused with "hallucinogenic alkaloids" from other flower sources such as *Datura*. These authors also mention that the vision serpents seen in ancient Maya iconography resemble those reported with *ayahuasca*, a tea containing powerful entheogenic chemicals (see Blainey 2013; Shanon 2002). While they ultimately dismiss this possibility because the standard botanical constituents of ayahuasca are found only in the Amazon rainforest, it is important to note that:

> It is entirely possible that *Banisteriopsis muricata* was used for this purpose, as its stems contain harmine and its leaves DMT. In other words, it is possible that an ayahuasca analog was made from just one plant... The vine is also found in the lowlands of southern Mexico (Selva Lacandona) and in Petén (Guatemala). (Rätsch 2005: 719, 89)

Future research is required to confirm or disprove this intriguing hypothesis.

Such notions add an entirely new dimension of ritual significance to royal court scenes like that on the painted vase K1453, described as a "palace drunken party" (Kerr 2013). Here we see one diminutive "dwarf" character drinking a yellowish-brown beverage while another dwarf holds a mirror in front of the principal noble figure (figure 9.4). In fact, "the only known scenes of drinking are those involving a dwarf... or other beings that appear to be supernatural (K1381, K4377)" (Houston et al. 2006: 127). Instead of presuming the liquid to be merely alcoholic, this is plausible evidence that the beverage being imbibed

FIGURE 9.4. *Detail of painted vessel K1453 showing a dwarf drinking a possibly entheogenic beverage while another holds a mirror (Kerr 2013).*

contained entheogenic substances potent enough to induce visions of what the Maya would have interpreted as Otherworldly dwarf beings. In fact, ethnohistoric sources directly infer that such "drunken parties" included entheogenic mushrooms. For instance, Franciscan friar Bernardino de Sahagún's sixteenth-century description of an Aztec "mushroom party":

> The first thing eaten at the gathering were certain black little mushrooms, which they call [*Teo-*]*nandcatl*, which inebriate and cause hallucinations, and even excite lust. These they ate before dawn, and they also drank cacao before dawn. The mushrooms they ate with honey, and when they began to get heated from them, they began to dance, and some sang and some wept, for now were they

FIGURE 9.5. *Sahagún's "imp."*

drunk from the mushrooms. And some cared not to sing, but would sit down
in their rooms, and stayed there pensive-like . . . Then when the drunkenness of
the mushrooms had passed, they spoke one with another about the visions that
they had seen. (Sahagún 1950, cited in Wasson and Wasson 1957: 223)

This same document includes an illustration of what Wasson and Wasson
(1957: 234, Figure 14, note 1) refer to as an "imp," a zoomorphic "gnome-like
creature" (de Borhegyi 1961: 503) depicted as rising out of a patch of entheogenic mushrooms.

Even though direct archaeological verification for the oral ingestion of
entheogens by the ancient Maya is more elusive, a graduate student alerted
me to objects held close to the face in some painted pottery scenes.[2] Normally,
these scenes are interpreted as Maya kings and nobles holding "small bouquets of flowers for sniffing" (Miller and Taube 1997: 89). However, in the
scenes depicted on K2914 and K2026 (figure 9.6), both of which contain
mirrors, it is clear that the noble persons sitting in front of the mirrors hold
these "bouquets" directly in front of their *open* mouths. From a common sense
standpoint, it is more likely that these objects were meant for the mouth (for

eating or licking) rather than for the nose, as sniffing or smelling is typically a closed-mouth activity. Whatever these handheld "bouquet" objects are (they do have a similar shape as the mushrooms seen in the Sahagún "imp" image mentioned above), the fact that they are being eaten or licked in front of an iron-ore mirror evokes significant connotations for the present discussion. More specifically, the mirror + "bouquet"-eating scene in K2026 includes both a water-bird and a bunch of rabbits behind and under the mirror, the same animals that were noted earlier as commonly occurring in painted pottery scenes that involve mirrors. In the scene on K2914, a "dwarf" character emits "speech scrolls" as he sits below the mirror the noble is looking into (and perhaps using to communicate with the underworld dwarf). What follows is an attempt to provide rational explanations for these recurrent motifs of undersized humanoid and animal beings occurring alongside mirrors in ancient Maya iconography. In essence, I suggest that these are "visions" caused by consuming entheogens prior to looking into the mirror.

"DWARFS" AS EVIDENCE FOR RITUAL COMBINATIONS OF MIRRORS AND ENTHEOGENS

Considering the evidence reviewed above regarding both the use of iron-ore mirrors and the ingestion of entheogenic substances by the ancient Maya, a reasonable supposition would infer that these two ritually significant items were occasionally combined. In exploring this possibility, I point to mirrors in painted polychrome images, which require a support so that their surface can be gazed into. Thus, we find that other than those mirrors that are being held by the very personality that is looking into them (see Kerr #'s 505, 8652), mirrors are most often depicted as supported by either a ring-like stand (see Kerr #'s 625, 2026, 2711, 2914, 2929, 3203, 5233, 5418 [eccentric, non ring-like], 5764, 6315, 6437, 6666 [upright bar stand], 7265, 8790, and 8793), or by an assistant of some kind. Interestingly, this assistant is sometimes what Mayanists refer to as a "dwarf" character (see Kerr #'s 530, 764, 787, 1453 [dwarf], 1454, 1463, 1790, 2025, 2695, 4338, 5110 [dwarf], 6341, 7288, 8220, and 8926).

Regarding the identity of the dwarf characters in these polychrome scenes, they bear striking resemblance to "extremely rare" wooden figures sourced to the Maya area. Despite the temporal instability of wood artifacts, Mayanist experts surmise that one particularly well-preserved sculpture "may once have held a square mirror, and the figure's pose suggests a trance state associated with divination. His features and costume are neither entirely Maya nor Olmec but rather combine aspects of the two styles" (Fields and Reents-Budet

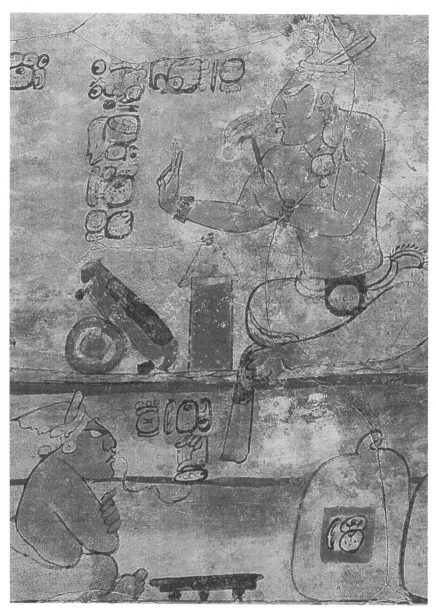

FIGURE 9.6. *Detail of painted vessel K2914, showing a noble apparently eating or licking "bouquets" or mushrooms (Kerr 2013).*

FIGURE 9.7. *"Extremely rare" wooden dwarf(?) sculpture in a "trance state" (Fields and Reents-Budet 2005: 106–107; http://www. metmuseum.org/toah/works-of-art/1979.206.1063).*

2005: 106). This peculiar wooden specimen (figure 9.7) has no solid provenance but is said to have been found in the border region between Guatemala and the Mexican state of Tabasco and is radiocarbon dated to AD 537 +/- 120 years (Ekholm 1964: 1–4). That this wooden sculpture once held a mirror in its lap is intimated by an almost identical specimen housed in the Princeton University Art Museum[3] (Catalogue Raisonnee: K4484a-c). A similar, yet badly decomposed wooden sculpture of comparable size (40 cm/15 in.) and style was found by a tourist in an unidentified Belizean cave and dates to the Early Classic period (Stuart and Housley 1999). Such "mirror bearer" figurines bear a striking resemblance to the dwarf figures (compare Figure 9.7 with Figure 9.4) depicted as holding mirrors on Maya painted pottery (see also Kerr's carved vessel K5110 [not shown]). It has been surmised that these wooden figures are the extraordinary remnants of what was once a widespread tradition of ancient Maya wooden dwarf mirror-holders (Houston et al. 2006: 196).

What was the significance of these "dwarf" characters in the context of ancient Maya sociopolitical life? A recent comprehensive survey of the "dwarf motif" on ancient Maya stone monuments concludes that "the data point to dwarfs as a symbol of liminality...its use on monuments celebrating the turning of calendric cycles reveals that the Maya viewed these points in time as liminal, both end and beginning" (Bacon 2007: 392). It is often assumed that these images of dwarfs indicate actual human members of the royal court born with a genetic disorder known as *achondroplasia*, possibly as a consequence of incest among elites (Miller 1985: 149–150; Miller and Taube 1997: 82). And yet in all the burials found at ancient Maya sites "no skeleton has been excavated that has proved conclusively to be that of a dwarf"[4] (Miller 1985: 141). This speculation about (otherwise archaeologically absent) dwarf members of the royal court must be reassessed in light of ethnographically established beliefs concerning dwarf spirit entities among the modern Maya.

Barbara Tedlock (1992: 455–456) reports that today "throughout the Mayan area, shamans, who practice as healers and dream interpreters, are selected for those roles by giants, dwarfs, and other tellurian deities who meet them when they are out walking in the hills and forests, visiting caves, or else while they are dreaming." The modern Maya association of a distinct race of "dwarf people" with intermediate spaces between the upper earth and the Underworld was also recounted by J. Eric S. Thompson (1970: 347). In addition, the Popul Vuh creation story mentions a "little man of the forest," just like forest-dwelling dwarf or "goblin" spirits that modern Maya call *alux* (Goetz et al. 1950: 84; Redfield and Villa Rojas 1962: 119–121; Schwartz 1990: 29, 72). Curiously, identical tales are told by cultures across the globe:

> Dwarfs are a type of elf or goblin found in the mythology and folklore of most of the world's societies. In general their description is of a humanoid shape but very small, looking old and wizened, with odd-shaped legs and feet. They usually inhabit caverns and underground palaces or dark forests, and are often associated with watercourses." (Rose 1996: 93)

The contention that the ancient Maya held similar beliefs about humanoid "little people" is supported archaeologically by the work of Stephen Houston, who has identified several glyphic references to these dwarfs. While Houston (1992: 528–530) maintains that dwarfs were real "individuals who performed a prominent, if carefully defined role at court," he also concludes that the ancients viewed these dwarfs much as the modern Maya do: as otherworldly beings "with close ties to the earth and to period ending rituals."

So it is established that the belief in small humanoid spirits is widespread and ancient in the Maya area. But what is the origin of this folklore tradition in the first place? A significant hint to this quandary is derived again from the ethnographic literature. As is noted by R. Gordon Wasson (1980: 37), an ethnomycologist who became famous for publicizing accounts of entheogenic mushroom ceremonies he attended in Oaxaca during the mid-twentieth century: "across linguistic lines throughout Mesoamerica the sacred mushrooms are called 'little children' by names that are always both affectionate and respectful." Wasson states that María Sabina, the Mazatec shaman who initiated him in his first mushroom ritual, told him that the "'children' . . . are the spirits of the mushroom." He also cites reports from colleagues who attended sessions where "'little people' answered the questions put to them by the shaman" (Wasson 1980: 134). Likewise, María Sabina claimed that she heard the voices of the "little children" or "little saint" spirits contained in the mushrooms, saying that they spoke to her when she was under their influence (Estrada 1981: 93; Schultes et al. 2001: 156–163). With modern use of entheogens in Mesoamerica, Wasson (1963: 184) commonly found that "in the accounts of the visions that the Indians see after they consume the sacred food—whether seeds or mushrooms or plant—there frequently figure *hombrecitos*, 'little men,' *mujercitas*, 'little women,' *duendes*, 'supernatural dwarfs.'" The details of these ethnographic testimonies evoke the ancient evidence, recalling that "the Maya mushroom stones are figures 28 to 38 cm high, with a standing man or animal represented below the cap of a mushroom" (Guzmán 2003: 318). Further, the Vico and Coto dictionaries prove that the Contact-period Maya had words for *xibalbaj okox* ("underworld mushroom"), also known as the *k'aizalah okox* ("lost-judgement mushroom") (Furst 1976: 76–79; Sharer 1994: 542), indicating their knowledge of mind-altering fungi associated with subterrestrial spirit-realms.

Interestingly, a specific category of visual hallucinations, associated with both psychiatric disorders and the ingestion of entheogens, entails small spiritual entities entering the field of perception, sometimes in an interactive manner:

Lilliputian hallucinations—named after the experiences of Jonathan Swift's character, Gulliver, with the little people of the island of Lilliput—are frequently given special attention in descriptive reports. The patients report seeing tiny but otherwise normally shaped people, animals and objects. Such diminutive figures, however, have been described in a variety of metabolic, neurologic, *drug-induced* and psychiatric conditions and appear to have no specific localizing or etiologic significance. (Cummings and Miller 1987: 48–49; emphasis added)

These same visual peculiarities were described by indigenous Siberians who had customarily used *Amanita muscaria* ("fly agaric") mushrooms in shamanistic rites: "Ancient Siberian peoples referred to 'fly-agaric men' and 'amanita girls' in connection with the lilliputian hallucinations said to occur in the context of *Amanita* intoxication" (Blom 2010: 305). But what scientists describe as Lilliputian "hallucinations" are interpreted by others as real encounters with little people and animals of abnormally small size. The conviction that these miniature people and animals are *not* "hallucinations" but rather the manifestation of normally hidden aspects of reality, is a cross-cultural phenomenon associated with mushroom intoxications documented ethnographically in China (Arora 2008), Papua New Guinea (Reay 1977: 59), and Britain (Letcher 2001: 157). In addition to dwarf characters, recall that mirrors are also often associated with fauna such as rabbits and water-birds in ancient Maya iconography.

Rick Strassman (2001: 39–40), a medical researcher who conducted studies on human-subjects with DMT (N,N-Dimethyltryptamine), an entheogen in the same chemical class as the psilocybin found in some mushrooms, reviews the subjective effects:

Perceptual or sensory effects often, but not always, are primary. Objects in our field of vision appear brighter or duller, *larger or smaller*, and seem to be shifting shape and melting . . . are softer or louder, harsher or gentler. We hear new rhythms in the wind. *Singing or mechanical sounds appear in a previously silent environment* . . . Our emotions overflow or dry up. Anxiety or fear, pleasure or relaxation, all feelings wax and wane, overpoweringly intense or frustratingly absent. At the extremes lie terror or ecstasy . . . Time collapses: in the blink of an eye, two hours pass. Or time expands: a minute contains a never-ending march of sensations and idea . . . *We experience others influencing our minds or bodies—in ways that are beneficial or frightening.*" (emphasis added)

Considering the evidence reviewed previously, the default assumption should be that the ancient Maya had an informed and reverent knowledge of psychoactive plants and mushrooms. One need only contemplate the quote above and imagine what powerful experiential effects would develop if an ancient

Maya person undergoing this type of altered state of consciousness stared intently into the polished surface of a mosaic iron-ore mirror. What is especially notable about Strassman's (2001: 185) findings is that at least half of the modern "volunteers 'made contact' with 'them,' . . . research subjects used expressions like 'entities,' 'beings,' 'aliens,' 'guides,' and 'helpers' to describe them." It is important for the purposes of this chapter to recognize that the physiological responses to these chemicals are universal for human beings. So even though interpretations can vary across cultures, the appearance of what seem like elves and other non-human entities is a common theme of entheogenic visions (see Tramacchi 2006). Thus, scrying with iron-ore mirrors while under the influence of such substances would most likely produce similar sensations of one's being in the company of strange otherworldly life forms. It has been demonstrated herein that the sensual effects of entheogenic substances offer a robust standpoint for explaining why humanoid "dwarfs" and animals such as water-birds and rabbits appear alongside mirrors in ancient Maya iconographic scenes.

CONCLUSIONS

With shamanistic techniques, an individual's experiential field is modified so that often they feel that they are achieving a personal consultation with a secluded realm of intelligent beings not perceived when that individual is in the normal waking state. In this way, I have sought to present a fresh contribution to the "archaeology of shamanism" debate. In short, mirrors yield material proof that a classic component of the shaman's toolkit was habitually used within the ancient Maya royal court.

My overall treatment of ancient Maya mirrors revolves around their function as vehicles for "divinatory scrying," wherein political elites are consistently associated with these objects in shamanistic contexts. This interpretation is deduced from multiple disciplinary perspectives (archaeological, iconographic, epigraphic, and ethnohistorical), converging around a new reading of ancient Maya worldview according to the *reflective surface complex* (Blainey 2007). The reflective surface complex is proposed as a model for expressing the ancient Maya's shamanistic belief that all shiny surfaces could be exploited as portable portals linking the mundane human world with an otherwise distant spiritual Otherworld. Mirrors were esteemed as valuable, high-status items that retained a religious (and I argue "shamanic") function among the early Maya across a vast expanse of both time and space. Although the precise role of shamanism among the ancient Maya will likely remain a topic of debate, it is clear that the "polished, gleaming" qualities of an iridescent mosaic surface, being both "dark and light at once" (Joyce 2002: 149), evoked liminal connotations for iron-ore mirrors. I assert that these affiliations with liminality, combined with their depiction as communicative devices in iconography, advocate an emic explanation of mirrors as instruments of shamanistic journeys to a spiritual Otherworld.

In this sense, the mirrors of ancient Mesoamerica are a poignant symbol of the endeavor of archaeology itself, in that we use the ancient cultures we study as mirrors to contemplate our own understanding of the human condition. While it strikes the modern scientific mind as ridiculous, the evidence demands that we consider how ancient Maya ontology differed from our modern beliefs about the nature of reality; indeed, when the ancient Maya experienced Lilliputian hallucinations/visions (perhaps after ingesting an entheogen), their reaction would most likely have been to consider the experience as revelatory of spirit entities that actually exist in some alternate dimension. I intend to promote/reject neither the modern skeptical interpretation nor the ancient believer's views regarding the existence of such legendary "little people." Rather, my aim is merely to identify the correlation between the cross-cultural precedents of Lilliputian hallucinations/visions occasioned by entheogenic substances and the ancient Maya artistic motif of mirrors associated with animals and "dwarfs." It must also be pointed out that this class of hallucinations/visions explains the "little" mushroom spirits reported ethnographically, as well as the animal and humanoid effigies that are carved into ancient mushroom stone sculptures.

Through the course of my research on these iron-ore mirrors, it has often puzzled me how the ideology of mirrors as devices for communication with Otherworldly beings was justified. Did the ancient Maya elites just have extremely active imaginations, inventing conversations with fantastical creatures out of thin air? To think that this is the case has never sat well with me. However, when we consider the fact that the ancient Maya were probably well acquainted with entheogenic plants and mushrooms alongside the known psychological impact of the chemicals these organisms contain, it is credible that such substances were used to augment mirror scrying practices (for a present-day example of such practices, see chapter 12, this volume). This proposition of the combined use of entheogens and mirror scrying renders explicable the iconographic images showing human actors communing with nonhuman entities on either side of an iron-ore mirror. In sum, this chapter has proposed an inductive scenario whereby ancient Maya mirror-gazing was in all likelihood a shamanistic practice combined with the use of entheogenic substances: looking into the luminescent face of the mirror, a person under the influence of an entheogen would experience shifts in sensual and existential awareness while simultaneously perceiving the presence of uncanny humanoid and animal beings.

NOTES

1. Despite the previous identification (Blainey 2007: 135–136) of the figure in K2929 as simply a jaguar on account of its spotted body markings, Karl Taube (personal communication, 2011) has pointed out that the large ear and fluffy tail are those of a rabbit (see Healy and Blainey 2011: 233). Thus, it is more prudent to designate these conflated qualities as those of a rabbit-jaguar *wahy* character.

2. Having read my BA thesis (Blainey 2005), L. Vinoski-Taylor (personal communication, 2008), a graduate student at the University of North Carolina, Wilmington, wrote me an email outlining her theory that these handheld fan-like objects were clumps of psilocybin mushrooms in the process of being licked or eaten.

3. http://artmuseum.princeton.edu/collections/objects/33418

4. The only potential example is the remains of a "short individual" found in Tikal Burial 24, but "his bones could not be salvaged; thus it was not determined whether he was a true dwarf" (Miller 1985: 141, 150).

REFERENCES

Arora, David. 2008. "Xiao Ren Ren: The "Little People" of Yunnan." *Economic Botany* 62 (3): 540–544. http://dx.doi.org/10.1007/s12231-008-9049-0.

Bacon, Wendy J. 2007. "The Dwarf Motif in Classic Maya Monumental Iconography: A Spatial Analysis." PhD diss., University of Pennsylvania, Philadelphia, PA.

Blainey, Marc. 2005. "Evidence for Ritual Use of Entheogens in Ancient Mesoamerica and the Implications for the Approach to Religion and Worldview." BA thesis, Department of Anthropology, University of Western Ontario. Wayeb: The European Association of Mayanists. www.wayeb.org/download/theses/blainey _2005.pdf.

Blainey, Marc. 2007. "Surfaces and Beyond: The Political, Ideological, and Economic Significance of Ancient Maya Iron-Ore Mirrors." MA thesis, Department of Anthropology, Trent University, Peterborough, Ontario. Wayeb: The European Association of Mayanists. http://www.wayeb.org/download/theses /blainey_2007.pdf.

Blainey, Marc. 2010. "Deciphering Ancient Maya Ethno-Metaphysics: Conventional Icons Signifying the 'King-as-Conduit' Complex." *Time and Mind* 3 (3): 267–289. http://dx.doi.org/10.2752/175169610X12754030955896.

Blainey, Marc. 2011. "An Anthropology of Third-Wave Mayans: Emic Rationales Behind New Age (Miss)Appropriations of Ancient Maya Calendrics and Symbology." In *Identity Crisis: Archaeology and Problems of Social Identity*, 42nd Annual Chacmool Conference Proceedings, 2009, edited by L. Amundsen, S. Pickering and G. Oetelaar, 352–365. Calgary: Chacmool Archaeological Association, University of Calgary.

Blainey, Marc. 2013. "A Ritual Key to Mystical Solutions: Ayahuasca Therapy, Secularism, and the Santo Daime Religion in Belgium." PhD diss., Department of Anthropology, Tulane University, New Orleans, LA.

Blom, Jan Dirk. 2010. *A Dictionary of Hallucinations*. New York: Springer. http://dx .doi.org/10.1007/978-1-4419-1223-7.

Calvin, Inga. 1997. "Where the Wayob Live: A Further Examination of Classic Maya Supernaturals." In *The Maya Vase Book*, vol. 5. ed. Justin Kerr, 868–883. New York: Kerr Associates.

Cano, Mirtha, and Nicholas Hellmuth. 2008. "Sacred Maya Flower: *Nymphaea ampla* Salisb." *FLAAR Reports.* http://www.wide-format-printers.org/FLAAR_report _covers/705177_Waterlili_Report.pdf.

Carod-Artal, F. J. 2011. "Alucinógenos en las culturas precolombinas mesoamericanas" [Hallucinogenic drugs in pre-Columbian Mesoamerican cultures]. *Neurología.*

Cummings, J. L., and B. L. Miller. 1987. "Visual Hallucinations. Clinical Occurrence and Use in Differential Diagnosis." *Western Journal of Medicine* 146 (1): 46–51.

Davis-Salazar, Karla L. 2003. "Late Classic Maya Water Management and Community Organization at Copan, Honduras." *Latin American Antiquity* 14 (3): 275–99. http://dx.doi.org/10.2307/3557560.

de Borhegyi, Carl, and Suzanne de Borhegyi-Forrest. 2013. "The Genesis of a Mushroom/Venus Religion in Mesoamerica." In *Entheogens and the Development of Culture: The Anthropology and Neurobiology of Ecstatic Experience*, ed. John A. Rush, 451–484. Berkeley, CA: North Atlantic Books.

de Borhegyi, Stephan F. 1961. "Miniature Mushroom Stones from Guatemala." *American Antiquity* 26 (4): 498–504. http://dx.doi.org/10.2307/278737.

de Borhegyi, Stephan F. 1963. "Pre-Columbian Pottery Mushrooms from Mesoamerica." *American Antiquity* 28 (3): 328–338. http://dx.doi.org/10.2307/278276.

de Smet, Peter A. G. M. 1985. *Ritual Enemas and Snuffs in the Americas*. Amsterdam, Netherlands: Centre for Latin American Research and Documentation.

de Smet, Peter A. G. M., and Nicholas M. Hellmuth. 1986. "A Multidisciplinary Approach to Ritual Enema Scenes on Ancient Maya Pottery." *Journal of Ethnopharmacology* 16 (2–3): 213–262. http://dx.doi.org/10.1016/0378-8741(86)90091-7.

Dobkin de Rios, Marlene. 1974. "The Influence of Psychotropic Flora and Fauna on Maya Religion [and Comments and Reply]." *Current Anthropology* 15 (2): 147–164. http://dx.doi.org/10.1086/201452.

Ekholm, Gordon F. 1964. *A Maya Sculpture in Wood*. New York: Museum of Primitive Art.

Eliade, Mircea. 1964. *Shamanism: Archaic Techniques of Ecstasy*. Princeton, NJ: Princeton University Press.

Emboden, W. A. 1981. "Transcultural Use of Narcotic Water Lilies in Ancient Egyptian and Maya Drug Ritual." *Journal of Ethnopharmacology* 3 (1): 39–83. http://dx.doi.org/10.1016/0378-8741(81)90013-1.

Emboden, W. A. 1982. "The Mushroom and the Water Lily: Literary and Pictorial Evidence for Nymphaea as a Ritual Psychotogen in Mesoamerica." *Journal of Ethnopharmacology* 5 (2): 139–148. http://dx.doi.org/10.1016/0378-8741(82)90039-3.

Erowid. 2006. "Psychoactive Lotus /Lily." http://www.erowid.org/plants/lotus/lotus.shtml.

Estrada, Alvaro. 1981. *María Sabina: Her Life and Chants*. Santa Barbara, CA: Ross-Erikson.

Fialko, Vilma. 2000. "El Espejo Del Entierro 49; Morfología y Texto Jeroglífico." In *El Sitio Maya De Topoxté: Investigaciones En Una Isla Del Lago Yaxhá, Petén, Guatemala*, edited by Wolfgang W. Wurster, 144–149. Mainz am Rhein: Verlag Phillip von Zabern.

Fields, Virginia M., and Dori Reents-Budet. 2005. *Lords of Creation: The Origins of Sacred Maya Kingship*. Los Angeles: Los Angeles County Museum of Art in association with Scala Publishers.

FLAAR (Foundation for Latin American Anthropological Research). 2008. "Plants Utilized by the Maya from Classic Times through Today." *FLAAR Reports*. http://www.wide-format-printers.org/FLAAR_report_covers/705182_Plants_utilized_by_the_mayan.pdf.

Forte, Robert. 1997. *Entheogens and the Future of Religion*. San Francisco: Council on Spiritual Practices.

Freidel, David A., Linda Schele, and Joy Parker. 1993. *Maya Cosmos: Three Thousand Years on the Shaman's Path*. New York, W.: Morrow.

Furst, Peter T. 1976. *Hallucinogens and Culture*. San Francisco: Chandler and Sharp.

Goetz, Delia, Sylvanus Griswold Morley, and Adrián Recinos. 1950. *Popol Vuh: The Sacred Book of the Ancient Quiché Maya*. Norman: University of Oklahoma Press.

Griffiths, Mark. 2010. *Search of the Sacred Flower*. New York: St. Martin's Press.

Grofe, Michael J. 2006. "Glyph Y and GII: The Mirror and the Child." Glyph Dwellers, Report 21. Maya Hieroglyphic Database Project, at the University of California, Davis. http://nas.ucdavis.edu/sites/nas.ucdavis.edu/files/attachments /R21.pdf.

Gronemeyer, Sven, and Barbara MacLeod. 2010. "What Could Happen in 2012: A Re-Analysis of the 13-Bak'tun Prophecy on Tortuguero Monument 6." Wayeb Notes 34. http://www.wayeb.org/notes/wayeb_notes0034.pdf.

Grube, Nikolai, and Werner Nahm. 1994. "A Census of Xibalba: A Complete Inventory of Way Characters on Maya Ceramics." In *The Maya Vase Book*, vol. 4. ed. Justin Kerr, 686–715. New York: Kerr Associates.

Guzmán, Gastón. 2003. "Fungi in the Maya Culture: Past, Present and Future." In *The Lowland Maya Area: Three Millennia at the Human–Wildland Interface*, ed. Arturo Gómez-Pompa, 315–325. Binghamton, NY: Food Products Press.

Healy, Paul, and Marc Blainey. 2011. "Ancient Maya Mosaic Mirrors: Function, Symbolism, and Meaning." *Ancient Mesoamerica* 22 (2): 229–244. http://dx.doi.org /10.1017/S0956536111000241.

Houston, Stephen D. 1992. "A Name Glyph for Classic Maya Dwarfs." In *The Maya Vase Book*, vol. 3. ed. Justin Kerr, 526–531. New York: Kerr Associates.

Houston, Stephen D., and David Stuart. 1989. *The Way Glyph: Evidence for "Co-Essences" among the Classic Maya*. Washington, DC: Center for Maya Research.

Houston, Stephen D., David Stuart, and Karl A. Taube. 2006. *The Memory of Bones: Body, Being, and Experience among the Classic Maya*. Austin: University of Texas Press.

Houston, Stephen, and Karl Taube. 2000. "An Archaeology of the Senses: Perception and Cultural Expression in Ancient Mesoamerica." *Cambridge Archaeological Journal* 10 (2): 261–294. http://dx.doi.org/10.1017/S095977430000010X.

Humphrey, Caroline. 2007. "Inside and Outside the Mirror: Mongolian Shamans' Mirrors as Instruments of Perspectivism." *Inner Asia* 9 (2): 173–195. http://dx.doi .org/10.1163/146481707793646557.

Hutson, Scott R. 1999. "Technoshamanism: Spiritual Healing in the Rave Subculture." *Popular Music and Society* 23 (3): 53–77. http://dx.doi.org/10.1080 /03007769908591745.

Joyce, Rosemary A. 2002. *The Languages of Archaeology*. Oxford: Blackwell. http:// dx.doi.org/10.1002/9780470693520.

Kehoe, Alice Beck. 2000. *Shamans and Religion: An Anthropological Exploration in Critical Thinking*. Prospect Heights, IL: Waveland Press.

Kerr, Justin. 2013. "Maya Vase Database: An Archive of Rollout Photographs." *FAMSI* (Foundation for the Advancement of Mesoamerican Studies). http://research.famsi.org/kerrmaya.html.

Klein, Cecelia F., Eulogio Guzmán, Elisa C. Mandell, and Maya Stanfield-Mazzi. 2002. "The Role of Shamanism in Mesoamerican Art: A Reassessment." *Current Anthropology* 43 (3): 383–419. http://dx.doi.org/10.1086/339529.

Letcher, Andy. 2001. "The Scouring of the Shire: Fairies, Trolls and Pixies in Eco-Protest Culture." *Folklore* 112 (2): 147–161. http://dx.doi.org/10.1080/0015587 0120082209.

Litvinskii, B. A. 1987. "Mirrors." In *The Encyclopedia of Religion*, ed. Mircea Eliade, 556–9. New York: Macmillan.

Macri, Martha J., and Matthew G. Looper. 2003. *The New Catalog of Maya Hieroglyphs: The Classic Period Inscriptions*. vol. 1. Norman: University of Oklahoma Press.

McGraw, John A. 2004. *Brain and Belief: An Exploration of the Human Soul*. Del Mar, CA: Aegis.

Miller, Mary Ellen, and Karl Taube. 1997. *An Illustrated Dictionary of the Gods and Symbols of Ancient Mexico and the Maya*. London: Thames and Hudson.

Miller, Virginia E. 1985. "The Dwarf Motif in Classic Maya Art." In *Fourth Palenque Round Table, 1980*, ed. Elizabeth P. Benson, 141–54. San Francisco: Pre-Columbian Art Research Institute.

Montgomery, John. 2002. *Dictionary of Maya Hieroglyphs*. New York: Hippocrene Books.

Narby, Jeremy, and Francis Huxley. 2001. *Shamans through Time*. New York: J. P. Tarcher/Putnam.

Nelson, Zachary, Barry Scheetz, Guillermo Mata Amado, and Antonio Prado. 2009. "Composite Mirrors of the Ancient Maya: Ostentatious Production and Precolumbian Fraud." *PARI Journal* 9 (4): 1–7.

Nielsen, Jesper. 2006. "The Queen's Mirrors: Interpreting the Iconography of Two Teotihuacan Style Mirrors from the Early Classic Margarita Tomb at Copan." *PARI Journal* 6 (4): 1–8.

Pearson, James L. 2002. *Shamanism and the Ancient Mind: A Cognitive Approach to Archaeology*. Walnut Creek, CA: AltaMira Press.

Price, Neil S. 2001. *The Archaeology of Shamanism*. London: Routledge.

Prufer, Keith M. 2005. "Shamans, Caves, and the Roles of Ritual Specialists in Maya Society." In *In the Maw of the Earth Monster*, ed. James E. Brady and Keith M. Prufer, 186–222. Austin: University of Texas Press.

Rätsch, Christian. 2005. *The Encyclopedia of Psychoactive Plants: Ethnopharmacology and Its Applications*. Rochester, VT: Park Street Press.

Reay, Marie. 1977. "Ritual Madness Observed: A Discarded Pattern of Fate in Papua New Guinea." *Journal of Pacific History* 12 (1): 55–79. http://dx.doi.org/10.1080 /00223347708572313.

Redfield, Robert, and Alfonso Villa Rojas. 1962. *Chan Kom: A Maya Village*. Chicago: University of Chicago Press.

Renfrew, Colin. 1986. "Varna and the Emergence of Wealth in Prehistoric Europe." In *The Social Life of Things: Commodities in Cultural Perspective*, ed. Arjun Appadurai, 141–168. New York: Cambridge University Press. http://dx.doi. org/10.1017/CBO9780511819582.007.

Rose, Carol. 1996. *Spirits, Fairies, Gnomes, and Goblins: An Encyclopedia of the Little People*. Santa Barbara, CA: ABC-CLIO.

Ruck, C. A., J. Bigwood, D. Staples, J. Ott, and R. G. Wasson. 1979. "Entheogens." *Journal of Psychedelic Drugs* 11 (1–2): 145–47.

Sahagún, Bernardino de. (Original work published 1590) 1950. *General History of the Things of New Spain: Florentine Codex*. Santa Fe: School of American Research.

Saunders, Nicholas J. 1988. "Anthropological Reflections on Archaeological Mirrors." In *Recent Studies in Pre-Columbian Archaeology*, edited by Nicholas J. Saunders and Olivier de Montmillin, 1–39. BAR International Series 421. Oxford: British Archaeological Reports.

Schele, Linda, and Jeffrey H. Miller. 1983. *The Mirror, the Rabbit, and the Bundle: "Accession" Expressions from the Classic Maya Inscriptions*. Washington, DC: Dumbarton Oaks Research Library and Collection.

Schultes, Richard Evans, Albert Hofmann, and Christian Rätsch. 2001. *Plants of the Gods: Their Sacred, Healing, and Hallucinogenic Powers*. Rochester, VT: Healing Arts Press.

Schwartz, Norman B. 1990. *Forest Society: A Social History of Petén, Guatemala*. Philadelphia: University of Pennsylvania Press.

Shanon, Benny. 2002. *The Antipodes of the Mind: Charting the Phenomenology of the Ayahuasca Experience*. Oxford: Oxford University Press.

Sharer, Robert J. 1994. *The Ancient Maya*. Stanford, CA: Stanford University Press.

Shook, Edwin M., and Alfred Vincent Kidder. 1952. *Mound E-III-3, Kaminaljuyu, Guatemala*. Washington: Carnegie Institution of Washington.

Strassman, Rick. 2001. *DMT: The Spirit Molecule: A Doctor's Revolutionary Research into the Biology of Near-Death and Mystical Experiences*. Rochester, VT: Park Street Press.

Stross, Brian, and Justin Kerr. 1990. "Notes on the Maya Vision Quest through Enema." In *The Maya Vase Book*, vol. 2. ed. Justin Kerr, 348–361. New York: Kerr Associates.

Stuart, George E., and R. A. Housley. 1999. *A Maya Wooden Figure from Belize: Una Figura Maya De Madera, Proveniente de Belice.* Washington, DC: Center for Maya Research.

Taube, Karl A. 1983. "The Teotihuacan Spider Woman." *Journal of Latin American Lore* 9 (2): 107–189.

Taube, Karl A. 1992. "The Iconography of Mirrors at Teotihuacan." In *Art, Ideology, and the City of Teotihuacan: A Symposium at Dumbarton Oaks, 8th and 9th October 1988,* edited by Janet C. Berlo, 169–204. Washington, DC: Dumbarton Oaks Research Library and Collections.

Tedlock, Barbara. 1992. "The Role of Dreams and Visionary Narratives in Mayan Cultural Survival." *Ethos (Berkeley, Calif.)* 20 (4): 453–476. http://dx.doi.org/10.1525/eth.1992.20.4.02a00030.

Thompson, J. Eric S. 1970. *Maya History and Religion.* Norman: University of Oklahoma Press.

Townsend, Joan B. 2001. "Modern Non-Traditional and Invented Shamanism." In *Shamanhood: Symbolism and Epic,* ed. Juha Pentikainen, 257–264. Budapest: Akademiai Kiado.

Tramacchi, Des. 2006. "Entheogens, Elves and Other Entities: Encountering the Spirits of Shamanic Plants and Substances." In *Popular Spiritualities: The Politics of Contemporary Enchantment,* edited by Lynne Hume and Kathleen McPhillips, 91–104. Burlington, VT: Ashgate.

Walsh, Roger N. 2007. *The World of Shamanism: New Views of an Ancient Tradition.* Woodbury, MN: Llewellyn Publications.

Walter, Mariko Namba, and Eva Jane Neumann Fridman. 2004. *Shamanism: An Encyclopedia of World Beliefs, Practices, and Culture.* Santa Barbara, CA: ABC-CLIO.

Wasson, R. Gordon. 1963. "Notes on the Present Status of Ololiuhqui and the Other Hallucinogens of Mexico." *Botanical Museum Leaflets, Harvard University* 20 (6): 161–212.

Wasson, R. Gordon. 1980. *The Wondrous Mushroom: Mycolatry in Mesoamerica.* New York: McGraw-Hill.

Wasson, Valentina Pavlovna, and R. Gordon Wasson. 1957. *Mushrooms, Russia, and History.* New York: Pantheon Books.

Winkelman, Michael. 2000. *Shamanism: The Neural Ecology of Consciousness and Healing.* Westport, CT: Bergin and Garvey.

Woodbury, Richard B. 1965. "Artifacts of the Guatemalan Highlands." In *Handbook of Middle American Indians: Archaeology of Southern Mesoamerica,* ed. Robert Wauchope and Gordon R. Willey, 163–180. Austin: University of Texas Press.

Znamenski, Andrei A. 2007. *The Beauty of the Primitive: Shamanism and the Western Imagination*. Oxford: Oxford University Press. http://dx.doi.org/10.1093/acprof:oso /9780195172317.001.0001.

10

Stones of Light

The Use of Crystals in Maya Divination

John J. McGraw

In this chapter, I aim to bridge the rewarding archaeological contributions from the earlier part of this volume with ethnographic work performed among contemporary Mayas, particularly as informed by ritual specialists. In considering the ritual use of iridescent quartz crystals by living Mayas, we may gain insights regarding their ancient ancestors' use of reflective pyrite mirrors. Though the direct historical approach poses its risks, it may yet lead to revelations. In addition to reviewing some of the ethnographic data, I also take pointers from Saunders, Taube, Blainey, and others in promoting a paradigm of functional analysis that situates archaeological and ethnographic data without projecting a particularism onto Maya contexts that may be inappropriate given the local epistemologies. Finally, I draw from cognitive science and religious studies to present a theory regarding the popularity and importance of quartz crystals in ritual activity, not only among the Maya, but worldwide. It is hoped that the analyses presented not only clarify ideas regarding quartz crystals in divination, but more broadly address the nature and importance of scrying in Maya religion.

Based on the presence of quartz crystals in a variety of Classic-period ritual assemblages, archaeologists have suggested their use in ancient Maya religion (Brady and Prufer 1999). Contemporary Maya peoples use crystals for ritual purposes as well; from the earliest ethnographies to present studies, crystals show up again and again as part of the ritual specialist's

DOI: 10.5876/9781607324089.c010

paraphernalia. As has been suggested throughout this volume in regards to the divinatory use of mirrors, crystals play an important role in a number of scrying practices. Indeed, one Maya ritual specialist I consulted even referred to crystals as "little mirrors."

Is the preponderance of these reflective objects and the characteristics attributed to them due to tradition alone or are there underlying cognitive tendencies that lead people to adopt crystals as ritual objects? Here it is suggested that the geological prevalence and unusual properties of quartz crystals (e.g., their combination of hardness and transparency, their prismatic characteristics) have led to their ritual usage in so many contexts, past and present. Such characteristics, as well as the "hyperregular" nature of crystals (for instance, quartz crystals always develop a hexagonal morphology), elicit inferences about the artificial creation of crystals; they stimulate cognitive representations of anthropomorphic design and perceived order in nature. Stumbling upon a bright, sharply faceted quartz crystal is akin to finding Paley's famous watch with its alluring insinuations of a watchmaker. These intuitive associations help to explain why crystals end up in ritual contexts more than any other stone.

Quartz is the most common mineral in the world. It is composed of crystallized silica; that is, highly regular, tetrahedral configurations of silicon and oxygen atoms. Combined, these two elements make up 75% of the Earth's crust. A report sponsored by the US Department of the Interior noted that "crystalline silica's pervasiveness in our technology is matched only by its abundance in nature. It's found in samples from every geological era and from every location around the globe" (Lujan and Ary 1992: 1).

Human use of quartz crystal dates back to the Paleolithic era (see UNESCO 1999). A cache of quartz tools discovered in Brazil have been dated to 25,000 years before present (Vázquez 1987: 8586). Fractured quartz yields glass-like shards fit for cutting and for use in projectile weapons. It would seem that its role in ritual, though, may have eclipsed its more utilitarian functions in many traditional societies. Crystals possess a host of attention-calling characteristics—transparency, hardness, reflectivity, natural facets—that have long drawn people to these dazzling minerals (figure 10.1). The discovery of quartz crystals in Mesoamerican archaeological sites should come as no surprise, given the stones' prevalence, but a more thorough review of their appearance in the archaeological record suggests a set of interesting hypotheses regarding their importance for ritual.

James Brady and Keith Prufer took on this task in a 1999 article in which they review numerous cave excavations where crystals have been discovered and

FIGURE 10.1. *Quartz crystal (photograph by John J. McGraw).*

hypothesize a relationship between cave ritual and quartz crystals. Drawing from ethnohistorical material to inform their hypothesis, Brady and Prufer associate caves and crystals with illness, curing, and divination. The authors find these connections strong enough to suggest that quartz be used as an index of ritual activity in Mesoamerican archaeological sites more generally.

In addition to the archaeological data, ample ethnohistorical evidence confirms the use of crystals for ritual purposes over long periods of time in a variety of Mesoamerican civilizations. Sahagún's (1963) Mexica informants described crystals (Classical Nahuatl, *tevilotl*) this way: "It is translucent, very transparent, clear. It is clear, very clear, exceedingly clear. Some are shaded, some are dense. They are cherished, esteemed, wonderful. They are precious, esteemed, venerated" (1963: 225). Similarly, Diego Durán (1994) described crystals as part of the tribute that Mexicas received from their subjects: "Large amounts of green stones, rock crystal, of carnelian, bloodstones, amber, besides many other types of precious stones that these people loved greatly. Their principal idolatry was the adoration of these stones" (1994: 203). Such were the sentiments regarding quartz crystals at the beginning of the colonial period. Later ethnographic accounts illustrate how such attitudes were little diminished by the five centuries that separate modern indigenous people from their early colonial forebears.

DIVINATORY SCRYING AND THE "REFLECTIVE SURFACE COMPLEX"

Why might crystals show up again and again in ritual contexts? For what purpose do ritual specialists collect and revere these stones? It seems that crystals are often associated with divinatory scrying, that is, obtaining occult information through flashes of light or images on any of a variety of reflective surfaces. Michael Winkelman discusses the kinds of objects used for scrying purposes:

> Some of the most important are mirrors, shiny reflective objects of stone or metal, bowls of water, bones, rocks, and crystals. Quartz crystals are frequently used as instruments for divining. Crystals may be used in diagnosis as a 'lens' for observing the patient's body, with the cause of disease being manifested within the crystal. Crystals may also be seen as the abode of spirits that could speak to the shaman and provide a diagnosis or other information. (2004: 81).

This generalization accurately characterizes material gleaned from ethnographies of many Mesoamerican groups, especially Maya peoples.

A review of Mesoamerican divinatory practices would include a great number of practices, generally varying from place to place. But two common "types" of divination include sortilege practices (which commonly involve the casting and selection of seeds) and scrying. Indeed, Karl Taube (1983: 120) identifies these two as "parts of a single divinatory complex in ancient Mesoamerica." Adapting Taube's notion of a "casting-scrying complex" in Mesoamerican ritual, Marc Blainey (Blainey 2007; Healy and Blainey 2011) posits a "reflective surface complex" among the ancient Maya. Blainey hypothesizes that all reflective surfaces, from still water to iron-ore mirrors to polished obsidian, possessed a functional similarity in regards to divinatory scrying.

In discussing a divinatory ritual, many researchers tend to overemphasize some particular technique, object, or prayer at the expense of more fundamental routines and relationships. Historical particularism, a mainstay assumption in both archaeology and anthropology, leads to fine-grained analyses that are often mute about larger patterns and generalizations. Nicholas Saunders (1988) addresses this theoretical issue, noting that "an overemphasis on technology can create a paradigmatic straightjacket, constantly generating ever more refined descriptive analyses whilst masking other potentially valuable insights into the possible function of mirrors and the wider notions surrounding reflected images" (1988: 2). Saunders indicates an affinity for functional analysis, at least in regards to the scrying complex which he, Taube, and Blainey find in their reviews of the archaeology of reflective objects. Moreover, this approach fits with the "cultural logic" (Fischer 2001) operative in the important notion of "co-essences," a pattern that indicates how ontological categories and divisions from European metaphysics are poor guides for understanding Maya "ethnometaphysics" (see Blainey 2010). A particular co-essence may link and regulate an aspect of time to an individual person, a type of plant, a feature of the environment, and weather phenomena (Monaghan 1998; Houston and Stuart 1989). In these cases, "family resemblance" may not easily show up in discernible surface features (a particular nose, tone of voice, or type of hair), but in qualities to be perceived in quite different ways, so that the claim that a person, a bird, a stone, and a lake are related is normative and intuitive for many Maya populations, past and present. In sum, surfaces are not to be trusted (see Warren 1998: 163–176), one must go deeper to identify a thing's characteristics and to determine its relationships to other things. Whether addressing the meaning of pyrite mirrors to the ancient Maya or the nature of co-essences among contemporary Maya peoples, specifying and categorizing objects and phenomena may require a set of ontological schemas different than the ones presently used by scholars.

Blainey's reflective surface complex suggests that the ancient Maya, like many people around the world, understood reflections to be a window into an alternate dimension. Ritual specialists paid special attention to unusual surfaces in order "to go through them" and access the "real" nature of things (see Kindl, chapter 12, and Taube, chapter 13, this volume). The dimension glimpsed in the flash of a dazzling crystal or in the smoky shadows of an obsidian mirror, is a plane of gods and ancestors, powerful forces behind and within—though sometimes hidden from—the world of men. It is this plane of reality that Maya ritual attends to and tries to influence. In reciting some set of prayers, enacting a ceremony, or making an offering, the ritual specialist attempts to sway, placate, or even coerce an invisible power to effect a change in the world. The scrying tool permits a window into this world and in the uncanny experience of finding a spark in a crystal or a face in the surface of the water, the diviner communicates with these powers.

ETHNOGRAPHIC EVIDENCE

Since 2007, I have consulted Tz'utujil, K'iche', and Kaqchikel ritual specialists, known as *ajq'ijaa'*.[1] Among my informants, divinatory crystals were called by three terms: *ilb'al* (K'iche', "tool for seeing"), *tz'atab'al* (Tz'utujil, "tool for observing or visualizing"), and *saqabaj* (Tz'utujil, "white or clear stone"). I often noticed quartz crystals when visiting these diviners. Crystals adorn their altars and divining tables, nestled in among the other ritual objects; commonly statues of Catholic saints, Maya deities, and archaeological pieces. More essentially, though, crystals are basic components of the diviners' bundles. Most commonly called *la vara* (Spanish, "staff or rod") or simply *tz'ite'*, this bundle serves as a badge of office. One *ajq'iij* described it to me as akin to a diploma. The ritual specialist typically receives her bundle on the day she is initiated as an *ajq'iij*. Composed of a set of red seeds called *tz'ite'*, *la vara* is an important ritual object and an index of authority. A species of coral tree (*Erythrina corallodendron*), *tz'ite'* has many magical connotations for highland Maya groups. Mixed in with the mass of seeds one almost always encounters a collection of quartz crystals. Sometimes these crystals play a direct role in the sortilege technique (see Tedlock 1992), other times they are placed to the side while the seeds are randomized and arranged during the ritual (figure 10.2).

One young woman I spoke with, who was training to become an *ajq'iij*, described crystals as conductors and transmitters. I asked her how they worked and she responded that they worked through the invocation of *Ahaaw* (Tz'utujil, "God"). She told me that the little crystals could show you the problem and its solution. When I asked her what one could see in the crystals, she simply described them as "little mirrors."

A Tz'utujil *ajq'iij* explained to me that the crystals, which he called *saqabaj*, brought particular energies to *la vara*. He also said that depending on the innate capacities of the diviner (as determined by his or her day of birth in the 260-day ritual calendar), he or she may be able to look into the crystal and perceive divinatory signs.

Another Tz'utujil *ajq'iij*, the one whose divining table is featured here (figure 10.3), possessed many crystals but said that he did not possess the capacity to look into them for divining purposes. He described that capacity as

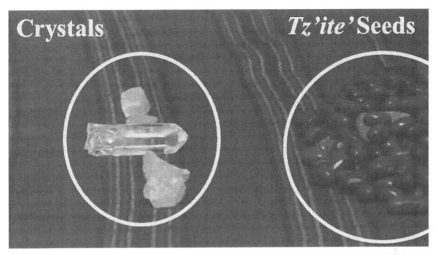

FIGURE 10.2. *Tz'ite' seeds and quartz crystals from a Tz'utujil diviner's bundle (photograph by John J. McGraw).*

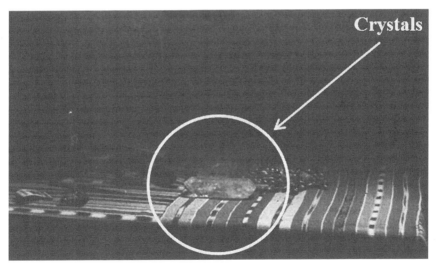

FIGURE 10.3. *Crystals on a Tz'utujil diviner's table (photograph by John J. McGraw).*

something that *ajq'ijaa'* possessed in the past (he claims that there were more powerful ones back then) but which has been lost over time.

Similarly, a high-ranking K'iche' *ajq'iij* was telling me about the various kinds of divination he knew of and mentioned that the highest type of diviner

FIGURE 10.4. *Crystal spheres used for diagnosis (photograph by John J. McGraw).*

used quartz crystals; he then qualified this by saying that such people are mentioned in old stories and are no longer around. Thus, two different specialists, from different ethnolinguistic groups, consider the use of crystals for divination to be a powerful means that has essentially been lost to time. In spite of these statements, I did manage to encounter two ritual specialists who still use quartz crystals as scrying instruments.

One of the people I interviewed, a young Tz'utujil man, is not an *ajq'iij*, but has become a respected healer. He possesses three crystals (more probably, glass), including two crystal spheres, which he uses in diagnosis. He claimed to have come across these objects while working in the hills. Like so many ritual specialists, then, these tools were set out for him to find by extraordinary or supernatural beings (see Brown 2000) (figure 10.4).

Another person, an elderly K'iche' woman in Quetzaltenango, also uses crystals as divining tools. I asked her about her crystals and she told me that she had obtained them in "holy places." She said she had visited some Maya altars, one near Cobán and another near Puerto Barrios, where she had found them. She said that not everyone has such luck, that she found them because of her particular gifts. I asked her what she saw inside the crystals and she tapped on her head and told me that she sees in her mind. I asked her if she saw visions. She responded that what she sees are not visions, but that God

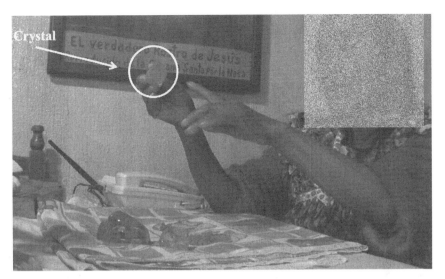

Crystal

FIGURE 10.5. *K'iche diviner (face occluded) using a crystal (photograph by John J. McGraw).*

has given her a light, a mental image that shows what is going on. She told me she could perceive if something was dead, in danger, or just about anything else (figure 10.5).

Crystals seem to be crucial components for ritual that my informants rely on in their work. Many of the *ajq'ijaa'* I consulted describe their *varas* as bundles of power, of "energy." Clearly, the *tz'ite'* seeds impart this power, as do the quartz crystals. In similar fashion, collecting these stones and keeping them on altars or divining tables add to the ritual power conveyed by the effigies, the archaeological pieces, the obsidian cores, and the Maya crosses. Though relatively few of the practitioners I have met use crystals for scrying, many kept crystals with their other ritual objects. Virtually all the specialists I have spoken with believe crystals to have special powers; they are capable of some kind of transference of "energy." A statement made by the Tedlocks' chief informant probably expresses this notion best; he compared the crystal to a radio capable of pulling in messages from far away (Tedlock 1993: 66; Tedlock 1992: 160). I encountered this radio metaphor myself. Of course, using a radio as a model for understanding the relationship of crystals to far off powers is necessarily a modern one. I would make the tentative hypothesis that what contemporary practitioners are describing as a "radio-like" capacity is similar to how ancient practitioners perceived their scrying tools vis-à-vis "the other world." Both refer to mediating technologies that transmit and receive information across vast spaces or different dimensions. The reflective surface complex is a radio-like association between transmitting powers that are far away or otherwise inaccessible and receiving artifacts, be they bowls of water, polished obsidian, pyrite mirrors, or quartz crystals.

THE THEORY OF AFFORDANCES IN
RELATION TO RITUAL OBJECTS

Archaeological, historical, and ethnographic evidence supports the use of quartz crystals by ritual specialists for the purposes of scrying over long periods of time. While Blainey's reflective surface complex is revealing in its emphasis on functional relationships rather than on specific objects, it is worthwhile to consider patterns of practice to determine if some objects lend themselves to ritual use more than others. In this case, for instance, even the functional complex selects for reflective surfaces rather than other kinds of surfaces. Reflectivity, then, is the essential characteristic; objects with greater reflectivity or, perhaps, objects that possess reflective characteristics as well as some additional characteristics that are favorable for ritual seem to be more prominent than other objects.

Are crystals on equal footing with polished obsidian, pyrite, bowls of water, and other reflective surfaces as just one among many potential tools for scrying? Comparing crystals to other reflective sources, a pattern emerges: obsidian is common in many places around the world but rarely has it enjoyed the ritual prominence that it does among the Maya. And water is everywhere but it is harder to document the various meanings and non-utilitarian uses that water has played for people throughout history. In contrast to more durable media, water is not so easily identifiable in material culture, at least insofar as it pertains to ritual. Before the modern age, the only common substance that possessed crystal clarity was water. Transparency, then, was a trait of liquids foremost and a few rare minerals, including the most abundant of them, quartz. In fact, the resemblance of quartz to water was so intuitive that Pliny the Elder (1962: 181) described how crystals were considered to be ice that had been frozen so long that it had lost the ability to melt. This observed resemblance came long before Pliny, though, as the word "crystal" comes from the ancient Greek word for "ice" (Hoad 1993). This "explanation" of the development of crystal persisted until the early modern period. Crystals have been linked with ritual since time immemorial and, in contrast to water, are tough enough to remain in the archaeological record indefinitely. Among ritual objects in general, crystals stand out as "special" ritual objects. Perhaps this is related to mundane properties, like their capacity to endure, but there are additional hypotheses worth examining.

As ritual objects the world over and through the ages, the popularity of crystals is matched only by bones and feathers, items that have direct associations with living beings and impressive capacities such as flight. A brief survey yielded numerous articles about the prehistoric use of crystals, seemingly for ritual purposes, in such diverse places as Ireland (Driscoll 2010), North America (Levi 1978; Furst 2006; Langenwalter 1980), South America (Vázquez 1987; Reichel-Dolmatoff 1979; Gil Sevillano 1997), and Australia (Rambo 1979). Crystals' prominent use among the Maya is merely an instance of a much larger pattern that connects people from across the globe and from deep history to the present moment. What is it about these objects that makes them so appealing for ritual?

Ann Taves (2013) adapts the language of "affordance" from Gibsonian psychology to suggest that certain objects "afford themselves" ritual associations and ascriptions of charisma more easily than others. James J. Gibson (1977) was a psychologist who sought to break down subject-object dualism, particularly as it related to the psychology of perception, and offered, instead, an ecological approach to perception: "The *affordances* of the environment are what it *offers* the animal, what it *provides* or *furnishes*, either for good or ill. The verb *to afford* is found in the dictionary, the noun *affordance* is not. I have made it up. I mean by it something that refers to both the environment and the animal in a way that no existing term does. It implies the complementarity of the animal and the environment" (1977: 127). Because humans alter and design the environments they live in, they engineer affordances into them as a matter of fact, as a basic design principle. A doorknob or handle affords grasping or pulling; indeed, this is why they were fashioned to possess the forms they do. Their "handy" shapes allow people to easily manipulate them. If they were built on a much smaller scale, for instance, they would lose most of the affordances they possess for grasping or pulling. Affordances illustrate how organisms and environments (be they natural or engineered) must be understood as coupled and coordinated systems rather than as separable, unrelated kinds.

Gibson (1979) employed a systems perspective to argue that humans do not perceive in an abstract, omniscient, or objective way but as the particular organisms we are; that is, we see *purposefully* and *selectively* and neither subject nor object take priority in this approach to perception (1979: 41). For example, note the parallel form that this stone face in the cave of Pech Merle shares with the head of a horse (figure 10.6). For humans who have observed horses,

FIGURE 10.6. *A horse (right) in Pech Merle cave is painted on a stone surface that resembles a horse's head (photograph by John J. McGraw).*

the resemblance is clear and immediate; it was because of this intuitive resemblance that upon seeing it, some prehistoric artist painted on a set of equine features to amplify the existing affordances. But those affordances were not "in the rock itself"; rather, they emerged from a human nervous system perceiving the stone from a particular vantage point. It then took an artist's ingenuity to enhance the affinity.

Taves harnesses Gibson's insights about affordances and Max Weber's notions regarding charisma to develop a theory regarding the "specialness" of particular objects in the context of religion. Weber's (1993: 3) ideas about charisma encompassed "natural objects, artifacts, animals, or persons." While the study of charisma in relation to authority figures has been carefully examined, the study of charisma in relation to "objects" remains relatively unexplored. Taves would have us take up this subject in order to better understand how special powers come to be attributed to objects. This approach is highly relevant to the present discussion; indeed, Maya ritual —past and present—would be difficult to conceive of without its elaborate ecology of special places, stones, and artifacts (Maxwell and García Ixmatá 2008).

Taves (2013) begins her assessment of the charisma of special objects with a set of programmatic questions: "Is there anything that reliably distinguishes the special from the ordinary across times and cultures? Is specialness simply a matter of discourse or cultural convention or is there something inherent either *in that which is set apart* or *in the way it is apprehended* or some complex combination thereof that is stable across cultures (and perhaps time)?" These inquiries can be usefully compared to Ferdinand de Saussure's (1959) discussion of the arbitrariness of sign relations. The word for something as fundamental to human beings as water is not to be predicted in any reliable way, except through the historical borrowings of languages, so that the Latin for water "aqua" and the English "water" cannot be traced to any cognitive relation that distinguishes them from any number of other arbitrary associations. But while arbitrariness is a driving principle in the association of particular sounds and particular objects, there are some regularities that emerge even in these sonic ecologies; onomatopoeia is a classic exception ("*woosh*, the car raced past me"). In such cases a pairing of the sound something makes, the displacement of air by a speeding car, comes to be mimicked in a word that resembles that sound. While arbitrariness may be central in linguistic forms, there is every reason to suppose that human visual systems (a far more robust sensory modality for us than hearing) seek out and attend to regularities in the environment in a much more active way, as seen in the case of the Pech Merle "horse."

Though we still lack a comprehensive cognitive theory of art, a few basic principles are difficult to deny. Thus, whether appreciating a visual form in Delhi or Delaware, a rich palette of colors is appreciated by all. And the universal lust for jewels and precious metals suggests a fundamental proclivity for the shiny and the dazzling. Moreover, while it is nearly impossible to find geometric rules and forms expressed in the ecosystems that humans inhabit (where can one find a straight line, an isosceles triangle, or a sharp angle out in the wild?) such regularities are quick indices of human action on a plastic medium. This effect is so reliable that stone implements dated to early hominid ancestors hundreds of thousands of years ago can be quickly discerned from the naturally fractured and water-shaped stones that may be encountered in the same assemblages or geological deposits. Debates emerge, no doubt, but even here at the dawn of human action on natural media, the regularities are quickly apprehended. There appears to be, then, a natural, intuitive visual capacity to perceive fine degrees of order in the chaotic forms of the natural world; to see "faces in the clouds," using the title of Guthrie's (1993) book on the subject. This has doubtlessly evolved over the millennia as a useful, if not essential, capacity for the interaction of humans vis-à-vis their ecosystems. But as with all human capacities, the most useful adaptations are also those that are most liable to become "hypersensitive" (see Barrett 2004: 24). The utility of perceiving order in nature, the capacity to perceive and finely discern "traces of intentionality" in highly complex visual scenes, has developed to such an extent that nature has favored the commission of so-called type 1 errors, "false positives," over and against errors of omission, type 2 errors, "false negatives" (see Hastie and Dawes 2010: 184–188). The sensitivity by which people may perceive paw prints on a trail heading toward the watering hole, and thus avoid dangerous predators, comes at the expense of seeing faces in the clouds, pyramids on the surface of Mars, and images in the broken surface of a pyrite mirror.

This capacity, this "hypersensitivity," to perceive order in underdetermined or "mildly chaotic" visual fields, when things are too far off or too out of focus to resolve clearly (sometimes called "pareidolia"), affords certain objects better functionality for scrying than others. Reflectivity is key, but irregular or broken reflectivity is preferable to stable, high fidelity reflectivity. Where scrying is still practiced, modern mirrors have yet to replace polished obsidian, crystals, or vessels of water. Why? Industrially produced mirrors are readily available and even, in many cases, much cheaper than a nice quartz crystal or a slab of polished obsidian. While tradition alone could explain the preference ("continue using the ritual objects that the ancestors used"), a more likely explanation can be drawn from the fact that a high fidelity reflective surface does not facilitate the mildly chaotic images necessary for the human visual system to resolve something that *is not immediately present*. Thus, the affordances of a nearly perfect reflective surface, like a modern mirror, promote their mundane functions in tending to one's appearance, but they do not afford a special function like scrying.

In addition to the affordances that quartz crystals offer in the way of producing an underdetermined visual scene fit for the generation of pareidolia in the diviner, they also possess a regularity of structure and form usually reserved for human artifacts. Compare a quartz crystal to an obsidian blade to get a sense of how much the former looks like the latter (Figure 10.7). The naturally geometric shapes of crystals—their hexagonal morphologies and straight-edged sides—exhibit formal regularities that stimulate inferences that they are worked artifacts. In short, "inferences of agency" are activated when one sees and handles a crystal.

William Paley's (1809) watchmaker analogy is apt here. In 1809, he made the compelling argument that design implies a designer. In his famous example, he compares the two kinds of inferences one might have upon encountering a stone or a watch. The stone is what it is and where it is because of a series of

FIGURE 10.7. *Obsidian blades and quartz crystals (photograph by John J. McGraw).*

natural (one might say "accidental") processes. But upon inspecting the watch, even if one had never beheld a watch, it would become clear that this was a human artifact and that human agency was responsible for its creation and its particular placement in the world. Indeed, similar inferences underlie the entire archaeological endeavor: if we had no such capacities, archaeologists would be incapable of distinguishing human artifacts from natural objects in the environment. Expanding these inferences to the world more generally, Paley aligns himself with the long tradition known as the "argument from design" which holds that the apparent systematicity of the world, its teleological orderliness, implies that it was designed to be just so. All the complex harmonies and orderly interactions that combine to make Earth a place fit for life were, to Paley, impossible to understand without referencing a designer. This human intuition for order, our quick apprehension for seeing purported complexity and design in a world that is more inchoate utterance than divine Word, also lies behind the sense of wonder experienced upon encountering a geometric kind—like a quartz crystal—in the wild.

And finally, besides possessing clarity with hardness and formal order, crystals simply stand out as wondrous things. Whatever the laws of beauty may be, the straight lines, sharp terminations, and gleaming facets of quartz crystals render them beautiful to most human eyes; they are attention-grabbing, at the very least. Looking into a crystal one may perceive the spectrum of light, as it is diffracted in the natural prism, or a cloudy wisp, where the crystallization remains imperfect, that leads to the kinds of fantastic musings one experiences when looking up at billowing clouds. Quartz crystals possess the affordances of being picked up, wondered over, and gazed upon; they afford speculation. They are, without doubt, the first lenses that human beings peered through, altering and transforming the shapes and sizes of things whose reflected light is channeled through them. All these curiosities combine to make them, in a word—uncanny—and thus do they appear to people when discovered in a river bed, automatically triggering many of the same inferences that Paley described in his watchmaker analogy.

Whereas ritual objects like relics receive their charisma through sympathetic magic, other objects, such as quartz crystals, are charismatic in and of themselves. Association to a ritual specialist may add to their power, but their initial power is not due to an association with a charismatic person. Sometimes, in fact, the causal arrows are reversed; that is, a person may be deemed charismatic because of the special objects he possesses and exhibits. In fact, religious charisma may best be understood by appreciating the complex "ritual ecology" of objects, places, and roles rather than any one of these separate from the others.

CONCLUSION

In a world of possibilities, humans seek out and make use of some objects, like quartz crystals, with a predictable regularity. Their functional affordances are not found in the objects themselves nor are they projected onto objects arbitrarily. Rather, the contingencies of human evolution have shaped a nervous system that seeks out and attends to some patterns, shapes, and visual properties more than others. Crystals stimulate many of these human sensitivities and additionally lead to inferences of being artificially created by purposeful agents. The activation of these uncanny representations have led quartz crystals to be selected as ritual objects more often than almost any other natural kind. As Brady and Prufer (1999) suggested in their paper, crystals may serve as an index of ritual activity in the archaeological record. Additional indices of use to archaeology may be discovered through careful analyses of contemporary ritual objects and settings (e.g., Maxwell and García Ixmatá 2008). The presence of crystals in numerous Maya archaeological sites as well as their use in contemporary Maya ritual suggests a historical continuity born partly from tradition. But whatever historical forces persevere are additionally buttressed by the affordances at play between crystals and the human visual system. Using archaeological analysis and cognitive anthropological principles, the affordances of certain objects for ritual can lead to testable hypotheses of aid to both fields, hypotheses that may permit us to peer—through a glass, darkly—into the lives and dreams of the ancestors.

NOTE

1. Brady and Prufer (1999) performed a useful survey of the literature and made the following list (which I have expanded) of the ritual use of quartz crystal among

modern Maya groups: Ch'orti'(Wisdom 1940: 345), Ixil (Colby and Colby 1981: 46), Jakaltek (La Farge and Byers 1931: 154), Kaqchikel (Tax 1946: 18; Tax and Hinshaw 1969: 94; Maxwell and Hill 2006: 21–22; Cuma Chávez 2005: 63), Q'eqchi' (Goubaud Carrera 1949: 39), K'iche' (Tedlock 1992; Tedlock 1993: 33; Schultze Jena 1947: 85; Bunzel 1959: 287; Freidel, Schele, and Parker 1993: 226; Hart 2008: 173; Itzep Chanchavac 2010; Tzoc Chanchavac, Tamup Aguilar, and Morales Pantó 2008: 15, 25; Molesky-Poz 2006: 76), Mam (Wagley 1949: 71; Oakes 1951a: 178; 1951b: 54), Mopan (Arvigo, Epstein, and Yaquinto 1994: 111–121), Q'anjob'al (Grollig 1959: 152; La Farge 1947: 182), Tz'utujil (Christenson 2001: 143; Douglas 1969: 139; Carlsen and Prechtel 1994: 97), and Yucatec (Hanks 1990: 246–251; Redfield and Villa Rojas 1962: 170; Freidel, Schele, and Parker 1993: 220–221; Hanks 2000: 197–217; Love 2004; Brinton 1883: 245; Astor-Aguilera 2010: 101).

REFERENCES

Arvigo, Rosita, Nadine Epstein, and Marilyn Yaquinto. 1994. *Sastun: My Apprenticeship with a Maya Healer.* 1st ed. San Francisco: Harper San Francisco.

Astor-Aguilera, Miguel Angel. 2010. *The Maya World of Communicating Objects: Quadripartite Crosses, Trees, and Stones.* Albuquerque: University of New Mexico Press.

Barrett, Justin L. 2004. *Why Would Anyone Believe in God?* Cognitive Science of Religion Series. Walnut Creek, CA: AltaMira Press.

Blainey, Marc. 2007. "Surfaces and Beyond: The Political, Ideological, and Economic Significance of Ancient Maya Iron-Ore Mirrors." MA thesis, Department of Anthropology, Trent University, Peterborough, Ontario.

Blainey, Marc. 2010. "Special Section: The Future of a Discipline: Considering the Ontological/Methodological Future of the Anthropology of Consciousness, Part II." *Anthropology of Consciousness* 21 (2): 113–138. http://dx.doi.org/10.1111/j.1556-3537 .2010.01025.x.

Brady, James E., and Keith M. Prufer. 1999. "Caves and Crystalmancy: Evidence for the Use of Crystals in Ancient Maya Religion." *Journal of Anthropological Research* 55 (1): 129-144.

Brinton, Daniel G. 1883. "The Folk-Lore of Yucatan." *Folk-Lore Journal* 1 (8): 244–256.

Brown, Linda A. 2000. "From Discard to Divination: Demarcating the Sacred Through the Collection and Curation of Discarded Objects." *Latin American Antiquity* 11 (4): 319–333. http://dx.doi.org/10.2307/972000.

Bunzel, Ruth Leah. (Original work published 1952) 1959. *Chichicastenango: A Guatamalan Village.* Seattle: University of Washington Press.

Carlsen, Robert S., and Martin Prechtel. 1994. "Walking on Two Legs: Shamanism in Santiago Atitlán, Guatemala." In *Ancient Traditions: Shamanism in Central Asia and the Americas*, ed. Gary Seaman and Jane S. Day, 77–112. Niwot: University Press of Colorado.

Christenson, Allen J. 2001. *Art and Society in a Highland Maya Community: The Altarpiece of Santiago Atitlán*. 1st ed. The Linda Schele series in Maya and pre-Columbian studies. Austin: University of Texas Press.

Colby, Benjamin N., and Lore M. Colby. 1981. *The Daykeeper: The Life and Discourse of an Ixil Diviner*. Cambridge, MA: Harvard University Press. http://dx.doi.org /10.4159/harvard.9780674283657.

Cuma Chávez, Baldomero. 2005. *Pensamiento Filosófico y Espiritualidad Maya*. Guatemala: Junajpu Editorial.

Douglas, Bill Gray. 1969. "Illness and Curing in Santiago Atitlan, a Tzutujil-Maya Community in the Southwestern Highlands of Guatemala." Thesis, Department of Anthropology, Stanford University, Palo Alto, CA.

Driscoll, Killian. 2010. "Understanding Quartz Technology in Early Prehistoric Ireland." Thesis, Archaeology, University College Dublin, Dublin.

Durán, Diego. 1994. *The History of the Indies of New Spain*. Translated by D. Heyden and F. Horcasitas. The Civilization of the American Indian Series. Norman: University of Oklahoma Press.

Fischer, Edward F. 2001. *Cultural Logics and Global Economies: Maya Identity in Thought and Practice*. Austin: University of Texas Press.

Freidel, David A., Linda Schele, and Joy Parker. 1993. *Maya Cosmos: Three Thousand Years on the Shaman's Path*. 1st ed. New York: William Morrow.

Furst, Peter T. 2006. *Rock Crystals and Peyote Dreams: Explorations in the Huichol Universe*. Salt Lake City: University of Utah Press.

Gibson, J. J. 1977. "The Theory of Affordances." In *Perceiving, Acting, and Knowing: Toward an Ecological Psychology*, ed. R. E. Shaw and J. Bransford, 127–43. Hillsdale, NJ: Lawrence Erlbaum Associates.

Gibson, J. J. 1979. *The Ecological Approach to Visual Perception*. Boston: Houghton Mifflin.

Gil Sevillano, J. 1997. "Lithic Tool Making by Amazonian Palaeoindians: A Case-Study on Materials Selection." *Journal of Materials Science Letters* 16 (6): 465–468. http://dx.doi.org/10.1023/A:1018560225861.

Goubaud Carrera, Antonio. 1949. *Notes on San Juan Chamelco*. Alta Verapaz: University of Chicago Microfilms.

Grollig, Francis Xavier. 1959. *San Miguel Acatán, Huehuetenango, Guatemala: A Modern Maya Village*. Bloomington: Indiana University.

Guthrie, Stewart. 1993. *Faces in the Clouds: A New Theory of Religion.* New York: Oxford University Press.

Hanks, William F. 1990. *Referential Practice: Language and Lived Space among the Maya.* Chicago: University of Chicago Press.

Hanks, William F. 2000. *Intertexts: Writings on Language, Utterance, and Context.* Lanham, MD: Rowman and Littlefield.

Hart, Thomas. 2008. *The Ancient Spirituality of the Modern Maya.* Albuquerque: University of New Mexico Press.

Hastie, Reid, and Robyn M. Dawes. 2010. *Rational Choice in an Uncertain World: The Psychology of Judgment and Decision Making.* Thousand Oaks, CA: Sage.

Healy, Paul F., and Marc G. Blainey. 2011. "Ancient Maya Mosaic Mirrors: Function, Symbolism, and Meaning." *Ancient Mesoamerica* 22: 229–244.

Hoad, T. F. 1993. *The Concise Oxford Dictionary of English Etymology.* Oxford: Oxford University Press.

Houston, Stephen D., and David Stuart. 1989. *The Way Glyph: Evidence for "Co-essences" among the Classic Maya.* Vol. 30. Washington, DC: Center for Maya Research.

Itzep Chanchavac, Rigoberto. 2010. El Origen de la Radiografía y los Rayos X Viene de la Cosmovisión Maya. Momostenango, Guatemala: Takiliben May.

La Farge, Oliver. 1947. *Santa Eulalia: The Religion of a Cuchumatán Indian Town,* University of Chicago Publications in Anthropology. Ethnological Series. Chicago: University of Chicago Press.

La Farge, Oliver, and Douglas Byers. 1931. *The Year Bearer's People.* Middle American Research Series Publication no. 3. New Orleans, LA: Department of Middle American Research, The Tulane University of Louisiana.

Langenwalter, Rebecca E. 1980. "A Possible Shaman's Cache from CA-Riv–102, Hemet, California." *Journal of California and Great Basin Anthropology* 2 (2): 233–244.

Levi, Jerome Meyer. 1978. "Wii'ipay: The Living Rocks—Ethnographic Notes on Crystal Magic Among Some California Yumans." *Journal of California Anthropology* 5 (1).

Love, Bruce. 2004. *Maya Shamanism Today: Connecting with the Cosmos in Rural Yucatán.* Lancaster, CA: Labyrinthos.

Lujan, M., Jr., and T. S. Ary. 1992. *Crystalline Silica Primer.* Washington, DC: Bureau of Mines Staff, Branch of Industrial Minerals U.S. Department of the Interior Special Publication.

Maxwell, Judith M., and Ajpub' Pablo García Ixmatá. 2008. *Power in Places: Investigating the Sacred Landscape of Iximche', Guatemala.* Mexico City: FAMSI.

Maxwell, Judith M., and Robert M. Hill. 2006. *Kaqchikel Chronicles: The Definitive Edition.* 1st ed. Austin: University of Texas Press.

Molesky-Poz, Jean. 2006. *Contemporary Maya Spirituality: The Ancient Ways Are Not Lost.* Austin: University of Texas Press.

Monaghan, John. 1998. "The Person, Destiny, and the Construction of Difference in Mesoamerica." *RES: Anthropology and Aesthetics* 33: 137–146.

Oakes, Maud. 1951a. *Beyond the Windy Place.* New York: Farrar, Straus and Young.

Oakes, Maud. 1951b. *The Two Crosses of Todos Santos: Survivals of Mayan Religious Ritual.* Princeton, NJ: Princeton University Press.

Paley, William. 1809. *Natural Theology.* 12th ed. London: J. Faulder.

Pliny. 1962. *Natural History.* Translated by D. E. Eichholz. Vol. 10. Loeb Classical Library. Cambridge, MA: Harvard University Press.

Rambo, A. T. 1979. "A Note on Stone Tool Use by the Orang Asli (Aborigines) of Peninsular Malaysia." *Asian Perspective* 22 (2): 113–119.

Redfield, Robert, and Alfonso Villa Rojas. 1962. *Chan Kom, a Maya Village.* Chicago: University of Chicago Press.

Reichel-Dolmatoff, Gerardo. 1979. "Desana Shamans' Rock Crystals and the Hexagonal Universe." *Journal of Latin American Lore* 5 (1): 117–128.

Sahagún, Bernardino de. 1963. *General History of the Things of New Spain, Book 11: Earthly Things.* Trans. A. J. O. Anderson and C. E. Dibble. Salt Lake City: University of Utah Press.

Saunders, Nicholas J. 1988. "Chatoyer: Anthropological Reflections on Archaeological Mirrors." In *Recent Studies in Pre-Columbian Archaeology*, ed. N. J. Saunders and O. D. Montmollin, 1–40. Oxford: British Archaeological Reports.

Saussure, Ferdinand de. 1959. *Course in General Linguistics.* New York: Philosophical Library.

Schultze Jena, Leonhard. 1947. *La vida y las creencias de los indígenas quichés de Guatemala.* Translated by A. G. Carrera and H. Sapper. Guatemala: Editorial del Ministerio de Educacion Publica.

Taube, Karl A. 1983. "The Teotihuacán Spider Woman." *Journal of Latin American Lore* 9 (2): 107–189.

Taves, Ann. 2013. "Charisma, Magic, and Other Non-Ordinary Powers: Special Affordances and the Study of Religion." In *Mental Culture: Classical Social Theory and Cognitive Science of Religion*, edited by D. Xygalatas and W. W. McCorkle, 80–97. London: Equinox.

Tax, Sol. 1946. *The Towns of Lake Atitlan.* Chicago: University of Chicago Microfilms.

Tax, Sol, and Robert Hinshaw. 1969. "The Maya of Midwestern Guatemala." In *Handbook of Middle American Indians*, ed. E. Z. Vogt, 69–100. Austin: University of Texas Press.

Tedlock, Barbara. 1992. *Time and the Highland Maya*. Rev. ed. Albuquerque: University of New Mexico Press.

Tedlock, Dennis. 1993. *Breath on the Mirror: Mythic Voices and Visions of the Living Maya*. 1st ed. San Francisco: Harper San Francisco.

Tzoc Chanchavac, Isabel, Felipe Tamup Aguilar, and Elvira Morales Pantó. 2008. *Enfermedades o Consecuencias?* Guatemala: Médicos Descalzos.

UNESCO. 1999. *The Peking Man World Heritage Site at Zhoukoudian*. Available from http://www.unesco.org/ext/field/beijing/whc/pkm-site.htm.

Vázquez, Juan Adolfo. 1987. "South American Indian Religions: Mythic Themes." In *Encyclopedia of Religion*, ed. L. Jones, M. Eliade, and C. J. Adams, 499–506. Detroit: Macmillan Reference.

Warren, Kay B. 1998. *Indigenous Movements and Their Critics: Pan-Maya Activism in Guatemala*. Princeton: Princeton University Press.

Wagley, Charles. 1949. *The Social and Religious Life of a Guatemalan Village*, Memoir Series of the American Anthropological Association, no. 7. Menasha, WI: American Anthropological Association.

Weber, Max. 1993. *The Sociology of Religion*. Trans. E. Fischoff. Boston: Beacon Press.

Winkelman, Michael. 2004. "Divination." In *Shamanism: An Encyclopedia of World Beliefs, Practices, and Culture*, ed. M. N. Walter and E. J. N. Fridman, 78–82. Santa Barbara, CA: ABC-CLIO.

Wisdom, Charles. 1940. *The Chorti Indians of Guatemala*. The University of Chicago Publications in Anthroplogy, Ethnological Series. Chicago: The University of Chicago Press.

11

Generally considered a pan-Mesoamerican cultural trait, mosaic iron-ore mirrors have most commonly been interpreted as elite ritual and status items, typically construed as the personal, revered effects of the Maya nobility during the Late Classic and Postclassic periods (ca. AD 600–1500). Referred to as "mirrors" due to their unique shape, size, and highly-reflective surfaces, these composite objects are constructed as "plaques usually composed of a slate backing with a shiny surface made of polished pyrite or hematite polygons assembled in a mosaic pattern" (Healy and Blainey 2011: 229). Where decorative applications exist (such as stucco, paint, or incised carvings), they tend to appear on the exterior surface of the slate backing, or "mirror back." Most examples have drill holes that penetrate cleanly through the surface at one or more edges, which may suggest that these mirrors, while in use, were suspended (perhaps around the neck or back) from clothing, or on a wooden frame as often depicted in Maya art (see chapter 1, this volume). Archaeologically, it is usually the slate backing that remains preserved in the record. Complete mirrors or partial fragments have been recovered from several primary archaeological contexts but are perhaps best known from their iconographic representation in various artistic media, especially ceramics. Recent research suggests that mirrors were not only elite ceremonial and status items, but likely served primarily as instruments of "religious divination and prognostication" used during scrying rites

Reflecting on Exchange

Ancient Maya Mirrors Beyond the Southeast Periphery

CARRIE L. DENNETT AND
MARC G. BLAINEY

DOI: 10.5876/9781607324089.c011

to communicate with the spiritual Otherworld (Healy and Blainey 2011: 229; see also Saunders 2003: 19, 22).

Curiously, these supposedly revered elite items have been noted in several regions of lower Central America,[1] far beyond the Mesoamerican southeast periphery, or frontier border. First defined in the 1940s by Kirchhoff (1981) and developed later by Willey (1966), the concept of the Mesoamerican southeast periphery distinguishes between the supposed "high cultures" of Mesoamerica to the north (characterized by advanced agriculture, monumental architecture, hieroglyphic writing, ritualized captive and autosacrifice, a pantheon of supernatural deities, and a traveling merchant class) and the "lower cultures" of lower Central America (characterized as loosely scattered, marginal, and seemingly backward groups). Traditionally, many—if not all—cultural advances witnessed in the archaeological record of lower Central America were interpreted as the result of either direct diffusion of material goods, external cultural influence, or actual population in-migrations by Mesoamerican groups from the north or Andean groups from the south (e.g., Baudez and Coe 1966; Snarskis 2003; Stone and Balser 1965). More recently, however, research has begun to model a cultural "Isthmo-Colombian area" that extends southward from northeast Honduras to northern Colombia and western Venezuela. The most current definition rests on the existence of shared linguistic traits (with all groups speaking some form of the broader Chibchan language family), shared genetic traits, evidence for stable, long-term occupation leading back several millennia in some regions, interregional material culture affinities, and shared worldviews—all of which are distinctly non-Mesoamerican (Hoopes and Fonseca 2003: 50–52; see also Cooke 2005: 161, Dennett 2007: 78; Healy and Dennett 2006; Hoopes 2005: 10–14). Importantly, this more balanced approach views the Mesoamerican southeast periphery as a mutually recognized geocultural interface between two unique pre-Hispanic culture areas.

Assuming that these mirrors were important personal effects of the Maya nobility, it seems odd to find examples dating from the Classic period forward outside of Maya "territory." Although their occurrence is relatively infrequent, the placement of iron-ore mirrors in archaeological contexts beyond the southeast periphery raises the question: How and why did these supposedly revered items make their way to lower Central America?

In this chapter we explore several occurrences of ancient mirrors in lower Central America, examining the artifacts themselves and the locations where they have been found, including a brief discussion of whether the example in question is a "real" Maya mirror or whether it is a locally produced "knockoff." We then explore traditional explanations that typically focus on arguments

CARRIE L. DENNETT AND MARC G. BLAINEY

of economic trade and/or the acquisition of foreign exotics for status elevation among emerging chiefly leaders beyond the southeast periphery of the
Maya area. When subjected to scrutiny, we find that these simplistic, "for
profit" arguments fail to adequately explain why the Maya elite might so easily
part with such ritual/status items. Finally, we present alternative interpretive
models to explain the presence of iron-ore mirrors in lower Central America.
These include concepts of sociopolitical emulation and developing "peer elite"
relationships that involve reciprocity in the form of "gifting."

IRON-ORE MIRRORS IN LOWER CENTRAL AMERICA

Iron-ore mirrors have been found in several archaeological regions of lower
Central America including northeast Honduras, the Atlantic Watershed
of Costa Rica, and the southern (Costa Rican) portion of Greater Nicoya.[2]
Unfortunately, professional publications discussing these finds in any archeological or interpretive detail are typically dated and disparate. Here we describe
and discuss several examples from these geographical and cultural regions
(figure 11.1) in order to shed new light on this enigmatic topic.

NORTHEAST HONDURAS

In the 1930s, fragments of two slate mirror backs were recovered from the
site of Wankybila in the vicinity of the Rio Patuca (Strong 1935), located on
the Caribbean mainland of northeast Honduras. These pieces are currently
housed at the Smithsonian Institution, and were analyzed there by Dennett in
2006. Associated artifacts, particularly ceramic types, suggest that the site had
an occupation spanning AD 500–1250 (Healy 1993; Lara Pinto 2006; Strong
1935). One of these mirror-back fragments exhibited a "rusty" residue of pitch,
a sticky vegetal resin, or some other adhesive used to hold the iron-ore mosaic
pieces onto the slate backing. Another specimen shows carved iconic motifs
and one fragment also displayed typical dual suspension holes. These features
suggest Mesoamerican-like production techniques, and a line drawing of a
main "scene" carved into one fragment helps better visualize and interpret the
weathered iconography (figure 11.2).[3]

Figure 11.2 exhibits a "quasi-Maya" carving style. Certain features (i.e., the
style of execution, shape, etc.) of the nostrils and lips of each figure resemble
examples from the recently discovered San Bartolo murals (Saturno 2006)
and certain eccentric flints from the Maya area (Miller 1999: 228). Other
aspects, however, conflict with conventional Classic Maya iconographic

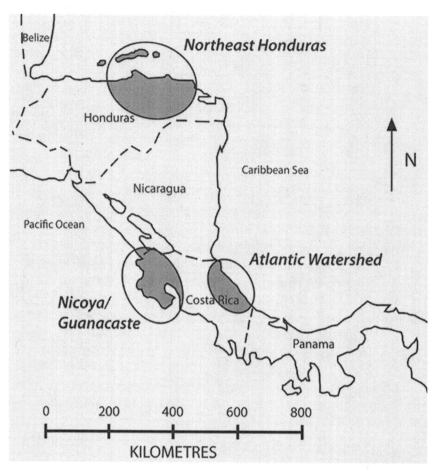

FIGURE 11.1. *Archaeological regions and subregions of lower Central America discussed in the text (map by Carrie Dennett).*

aesthetics, especially the sharpness of both figures' noses (Blainey 2007; see also Reents-Budet 1994), and it is uncertain whether the images actually represent Maya persons. It is possible that these fragments, which lack chronological control, may yield from a time period that predates Classic Maya artistic conventions (Blainey 2007: 49–50). The fragments themselves do share a material and technological resemblance to slate mirror backs from the site of Guácimo, in the Atlantic Watershed of Costa Rica (see below). It is perhaps more likely, however, that the piece is a local knockoff, or imitation, especially given the overall lack of stylistic fit between the Wankybila

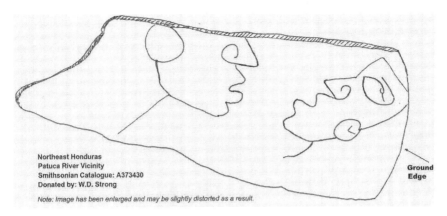

Northeast Honduras
Patuca River Vicinity
Smithsonian Catalogue: A373430
Donated by: W.D. Strong

Ground
Edge

Note: Image has been enlarged and may be slightly distorted as a result.

FIGURE II.2. *Line drawing of an incised slate mirror-back fragment from the site of Wankybila, northeast Honduras (drawing by Carrie Dennett).*

mirror fragments and examples known from the Maya region—a possibility that is discussed in greater detail below.

ATLANTIC WATERSHED OF COSTA RICA

The Atlantic Watershed of Costa Rica is, archaeologically speaking, a fairly well-known region in lower Central America due to the monumental construction of the "Linea Vieja," or "Old Line," by the United Fruit Company and the government of Costa Rica. A now defunct railway track (whose construction began in the late 1800s), the Linea Vieja opened up both the jungle and buried artifacts to workers, locals, and archaeologists (Jones and Morrison 1952; Snarskis 2003). Two sites along the line that stand out for their iron-ore mirror remains are La Fortuna and Guácimo, each of which is discussed below.

La Fortuna

Located in the San Carlos region of the central Atlantic Watershed, the site of La Fortuna is a pre-Hispanic cemetery that was victim to extensive looting by "huaqueros," or pothunters, before any archaeological work was ever conducted there. In fact it was a looter, Ricarte Moreno, who made the site known to the public with the discovery of an exquisitely incised Maya mirror back (Stone and Balser 1965: 311) (figure 11.3). This mirror is currently on display at the Museo Nacional de Costa Rica. In the early 1960s, Doris Stone and Carlos Balser recovered several mirror backs and/or fragments, some still bearing intact mosaic pyrite pieces, from six excavated burials. Notably, no grave

FIGURE 11.3. *Incised mirror back from the site of La Fortuna, located in the Atlantic Watershed of Costa Rica (from Schmidt, de la Garza, and Nalda 1998: 628, Catalogue No. 433; reproduction permission courtesy of, and photography by, Juan Vicente Guerrero Miranda).*

contained more than one mirror. Among the examples were a plain, undecorated mirror back, fragments bearing rare remnants of a wooden frame, a large mirror back fragment with remnants of red and white stucco, a necklace made of 10 reworked fragments, and the complete incised mirror initially found and made public by the looter mentioned above (Stone and Balser 1965: 313).

Associated artifacts and stylistic interpretation of the glyphs date the production and interment of the complete mirror to roughly AD 400–500 (Baudez and Coe 1966; Mora-Marín 2005; Stone and Balser 1965). The incised glyphs are present in two vertical columns, are Classic Maya in form and execution, and the piece was produced in southern Mesoamerica, most likely somewhere in the Peten region of Guatemala. In the 1960s, J. Eric S. Thompson initially characterized the columns of glyphs as being Early Classic Maya from the southern lowlands, especially similar to early glyphs at Tikal and Uaxuctun, but presented in a "nonsensical" arrangement. Thompson further suggested that the carvings were almost certainly merely decorative, recalling a corpus of individual glyphs that were often used to decorate Maya pottery (for Thompson's complete synopsis see Stone and Balser 1965: 316). Of interest, since Thompson's comments on the possible origin of the glyph styles, archaeologists discovered the Tikal Ballcourt Marker, which is self-dated to AD 416 and shares similar calligraphic style to the glyphs on the La Fortuna mirror back (Hoopes 2005: 20; Mora-Marín 2005).

Despite significant advances in our understanding of the Classic Maya writing system, recent attempts at deciphering the carved mirror back fail to provide any agreed-upon translation. For example, Juan Vicente Guerrero Miranda (Schmidt, de la Garza, and Nalda 1998: 628, Catalogue No. 433) suggests that the La Fortuna mirror "briefly tells of the ascent to power of a king: 'During the month of yaxkin during the month of [?] the king or lord [?], ninth successor of his lineage, tied the royal band for himself [crowned himself].'" Epigrapher Alexandre Tokovinine of Harvard University (personal communication 2007), on the other hand, suggests there is nothing certain with the text except that the structure appears to be a single, continuous column, which contradicts traditional dual-column text structure used by ancient Maya scribes. While the glyphs do appear to translate to some degree, they are not clearly traditional Classic Maya script. Forty years after Thompson's initial attempt to decipher the La Fortuna mirror back it seems that the only true consensus reached is that this piece was likely a "real" mirror produced by Maya artisans.

Guácimo

Located near the modern area of El Tres de Guácimo along the Linea Vieja in the central Atlantic Watershed, Stone and Balser (1965: 317) also excavated graves at the pre-Hispanic cemetery site of Guácimo. Again, each grave contained a single mirror or mirror fragment and all but one example had some of the pyrite mosaic pieces still intact. Other examples had traces of red and white stucco covering the exterior (rear) surface, and one particular specimen still had stone shims attached to the surface that were used to level the pyrite mosaic surface. Of the seven examples discovered at the site in the early 1960s,

FIGURE II.4. *Incised mirror backs from the site of Guácimo, located in the Atlantic Watershed of Costa Rica (from Stone and Balser 1965: 320–321, Figures 21 and 22, respectively; reproduced by permission of the Society for American Archaeology from American Antiquity 30 [3], 1965).*

two were discovered by looters. These were elaborately incised mirror backs and they are quite different than those from the site of La Fortuna. Stone and Balser (1965: 320–21) initially suggested that the carved iconography exhibited Classic Veracruz or Tlaxcalan stylistic motifs from Mexico (see figure II.4). However, several mirror backs carved in a similar style are known to have been produced at highland Maya sites such as Kaminalyuju and Zacaleu, Guatemala, suggesting that these motifs are likely native to the Maya subarea (Woodbury and Trik 1953). Interestingly, also possibly associated with this mirror were artifacts of gold and *tumbaga* (an alloy of primarily gold and copper ore; made to resemble gold through the process of depletion gilding) from Panama and Colombia, suggesting an extensive network of social connections centered at this site. These metal artifacts, local ceramic types associated with the Guácimo graves, and the style of incised glyphs on some of the mirror fragments suggest a date of roughly AD 420–520 for the mirror backs (Snarskis 2003).

All the complete and fragmented mirror backs from both La Fortuna and Guácimo burials appear to be "real" Maya imports to the site. Notably, although these mirrors have been judged to be actual imports from the Maya subregion of Mesoamerica, most have non-traditional iconographic programs. One particularly poignant example is the incised mirror back from La Fortuna.

The fact that the glyphs, like the mirror itself, are also largely "real" but do not follow the logic of Maya hieroglyphic writing is of great interest and will be revisited in the final interpretation section below.

GREATER NICOYA

Greater Nicoya is an archaeological region comprising two distinct subregions: Pacific Nicaragua to the north and Nicoya-Guanacaste, Costa Rica, to the south. To date, a few mirrors and/or fragments have been recovered from the southern subregion of Nicoya-Guanacaste (Lange 1992: 121). Like their Atlantic Watershed counterparts, the fabrication style and glyphic programs, where they exist, suggest that they are also Classic Maya products. What is different, however, is that epigraphers seem to have met with somewhat more success in deciphering carved inscriptions from Greater Nicoya.

One important example of a well-preserved, incised mirror back with detailed hieroglyphic text was excavated from a cemetery in the Bagaces area of the Nicoya-Guanacaste subregion that dates, at least in part, to sometime during the late Tempisque (AD 300–500) to Bagaces (AD 500–800) periods in the local chronology (Stone 1977: 64–65). The mirror back exhibits a dual-column arrangement of glyphs set into a shallowly carved, low-relief rectangular setting, which is quite unique. Tatiana Proskouriakoff (cited in Stone 1977: 64) provided the first published attempt (known to us) to translate the glyphs, suggesting that the glyphs included "statements introduced by the so-called 'Jog' (rodent) glyph (B1 and A4). The first statement concerns the birth of 'Turtle?' of 'Yaxchilan?'. It is followed by an expression 'in the reign of' (B5) and an unknown name. The last five glyphs contain familiar signs B6 (Cab or Caban—'earth, land, people?') and B7—the Sun-mask—but the meaning of the passage is unknown."

Two decades later, perhaps based on cues taken from Proskouriakoff's original reading, a more complete translation of the incised glyphs was provided by Juan Vicente Guerrero Miranda (Schmidt, de la Garza, and Nalda 1998: 629, Catalogue No. 434). He suggested in the 1990s that the glyphs read "[i]t was separating into two. The place seven-black-precious yellow was born from the carapace of the turtle uaxaktun [Uaxactún emblem glyph], he harvested from his tongue its sacred substance [he let blood] the son of the puma-turtle the protected son. Of jaguar divine sun eye or king of uak [emblem glyph]."

This same mirror-back, however, has recently been reanalyzed by epigrapher Alexandre Tokovinine and might represent a more accurate reading of many of the glyphs as representing the names of ancient Maya rulers at specific sites. Tokovinine (personal communication 2007) suggests that glyph blocks A1–A5 inform of an act of *ch'ab'-ak'ab'* (which means "do penance/ sacrifice" or "darkness") performed by the Sihyaj Chan lord named Sihyaj Chan Ahk for his god(s). There was a ruler named Sihyaj Chan Ahk ("Sky-born Turtle") at Piedras Negras (Martin and Grube 2000: 150). Tokovinine further tells us that "the inscription concludes (glyph blocks A7–B8) with saying that 'it' (presumably the mirror) is the gift of [*u-sib*] the Waka' lord [named] K'ihnich B'ahlam." Importantly, Waka' (known to archaeologists today as El Peru) is a site located in the west-central Peten region of northern Guatemala, so the "Waka' lord" reference might be to the king named K'ihnich B'ahlam at El Peru who ruled this site during the late seventh century AD. In addition, both Piedras Negras and El Peru are known to have been involved in political networks organized amidst the Calakmul sphere of influence during the late seventh century AD (Martin and Grube 2000: 109). With regard to the Nicoya-Guanacaste mirror, it is pertinent to the present discussion that this artifact refers directly to a gift exchange (probably involving the mirror itself) between two known Maya sites involving at least one confirmed historical figure of ancient Maya royalty.

Finally, Houston, Stuart, and Taube (2006: 60–67, Figure 2.2d) further suggest that glyph A4 on the Bagaces mirror back presents the glyphic expression *u-baah*, a Mayan term meaning 'his-[head, face, body, image, or self].' If this is correct, it is extremely important because it likely indicates that the mirror was meant to reflect either the image of the viewer or the image of an Otherworldly spiritual being that the viewer communicates with via the mirror (see chapter 9, this volume).

INTERPRETING MIRRORS IN LOWER CENTRAL AMERICA

In the section above we discussed several examples of mirrors and/or their component parts from northeast Honduras, as well as the Atlantic Watershed and Greater Nicoya regions of Costa Rica. Most examples are believed to be authentic Maya mirrors, produced in the Maya southern lowlands in the Classic period, with the possible exception of mirror-back fragments from northeast Honduras. Mirror backs from the Atlantic Watershed are especially curious because they demonstrate non-traditional Maya glyphic programs. Equally interesting are the mirror backs from the Nicoya-Guanacaste region of Costa Rica. Both of these examples exhibit glyphs that seem to contain some information about the mirror itself within the inscriptions.

Having established that pre-Hispanic Maya mirrors and/or mirror fragments exist in the archaeological record of lower Central America, how do we explain their presence there? We turn now to evaluate the evidence, providing new and revised explanations for the occurrence of "real" Maya iron-ore mirrors beyond the southeast periphery. The following section evaluates traditional aggrandizement and economic network models, and also explores new, more nuanced models based on concepts of emulation and developing "peer elite" relationships that involve reciprocity in the form of "gifting."

Most published discussions of iron-ore mirrors (not the glyphs that occur on many of them) occurring below the Mesoamerican southeast periphery are descriptive and anecdotal. Interpretation tends to be stunted, where it does exist, and generally involves "matter of fact" explanation (or expectation) with little rigorous assessment. Perhaps this is because researchers have felt that the situation was self-explanatory, though it seems more likely to be a means of bypassing one of the most difficult heuristic steps in archaeological interpretation—"the step from the material to the social" (Mol 2011: 74). Regardless, the two most frequently used explanations include hypotheses of aggrandizement and economics. Importantly, traditional explanations tend to utilize these two concepts in tandem. Aggrandizement is often stated as the goal in seeking Maya mirrors, while economic exchange is viewed as the means employed in procuring them. In keeping with this traditional theme, each of these explanatory approaches is discussed, in tandem, below.

AGGRANDIZEMENT HYPOTHESES

Some of the earliest explanations for foreign exotics in lower Central America generally envisage individuals pursuing leadership roles in ancient communities transitioning from heterarchical to hierarchical forms of social organization. In the 1970s, Mary Helms (1979, 1992) was arguably the first to fully develop the aggrandizement hypothesis—though she never explicitly called it this herself—to explain the presence of foreign exotics in the archaeological record of lower Central America. In fact, her seminal works guided most interpretation in the late twentieth century. The lure of this interpretative model has been strong and researchers continue to propose aggrandizement—although most recently with a combined focus on chiefly *and* shamanic ties—as the "prime mover" behind the acquisition of foreign exotics (e.g., see Snarskis 2003: 160–161).

Helms's (1992: 319–324) model of chiefly aggrandizement for lower Central America is seated in and organized around concepts of "distance" and cosmology. She interprets the ancients' universe as having both a horizontal and vertical axis. Thus, on the vertical axis, the most distant and significant realms include the most intangible places—the heavens and the underworld. Accordingly, on the horizontal axis, the most cosmologically significant and/or supernatural realms would then be the most distant and mysterious earthly places. For an individual seeking to establish authority it would be necessary to demonstrate access to, or contact with, these supernatural realms. Exotic artifacts from foreign places were thus imbued with cosmological significance, via the quality of "distance," and served as a tangible and direct means of demonstrating one's access to the earthly supernatural. Thus, when accessible, they served as symbols of rank, status, and cosmological connection. It then became the burden of the one contending for power and/or authority to personally travel to distant realms and seek out these tangible symbols.

Without a doubt, there was marked sociocultural development taking place by AD 500 throughout the areas of lower Central America discussed herein (Dennett 2007; Hoopes 2005: 21), which roughly coincides with the production and deposition dates related to the Maya mirrors discussed above. So it is possible to see why arguments for aggrandizement strategies among emerging chiefly leaders would be appealing to archaeologists attempting to explain the presence of exotic artifacts from far-flung and culturally distinct regions. There is difficulty, however, in advancing this argument as a complete explanation as to why Maya mirrors appear in lower Central America. The most problematic aspect of this potential explanation is that mirrors were—unlike more "common" exotic objects such as pottery or obsidian—individual-specific elite ritual and status items, imbued with the power to connect Maya nobles to their gods (Blainey 2007: 114). Because Maya nobility was based largely on ascribed status, there was less need to legitimate their rule to non-Maya visitors in aggrandizement schemes to secure their authority. Therefore, we find it implausible that Maya nobles would distribute these highly-valued objects to be used by culturally unconnected "outsiders" who might have arrived on a pilgrimage (or what-have-you) seeking Maya symbolic artifacts for their own status-elevation or aggrandizement purposes.

Economic Hypotheses

Economic networking models (sometimes in conjunction with revised aggrandizement hypotheses) have also been employed to explain the occurrence of Maya mirrors beyond the Mesoamerican southeast periphery. Since the 1960s, the movement of Maya mirrors into lower Central America has generally been interpreted as the result of economic trade, grouped (inappropriately) with other, more abundant, commodities such as pottery or obsidian. For example, Stone and Balser (1965) state plainly that the mirrors found in the Atlantic Watershed of Costa Rica were "important in commerce," intimating that mirrors were purchased through economic trade networks for use in local mortuary rites. As an extension, Baudez and Coe (1966) state that the movement of the La Fortuna mirror from its place of production in the Peten region of Guatemala to the Atlantic Watershed of Costa Rica is indicative of "commercial contacts." We are of the belief, however, that ancient Maya iron-ore mirrors were not "just another trade good."

Alternative research, still economically based, may hold clues that provide one possible explanation. It has been suggested that many types of Maya status objects may have entered into circulation in lower Central America via traveling traders who profited from "the result of lowland Maya warfare and tomb desecration in the fourth century AD" (Snarskis 2003: 169). We find this suggestion both interesting and problematic, and implore a reasonable explanation as to why Maya conquerors would sell, for profit, such rare spoils of war.

The ancient Maya had a penchant for flaunting—or making absolutely visible to the public—sociopolitical power, authority, and prowess. Glyphic texts were perhaps the most important medium through which nobility asserted and displayed power (Johnston 2001: 375), and such texts are most commonly witnessed on inscribed stelae and monumental architecture, either professing the supreme authority of a ruling king or boasting the conquering of a rival king. With regard to tomb desecration and the "spoils of war" hypothesis, it may be possible that dispersing the mirrors was a viable method of removing them from the Maya sociopolitical domain. However, we believe it more likely that they would be destroyed as is often seen with the desecrated stelae and decimated royal architecture of conquered kings during the Classic period—perhaps ritually (or brutally) disposed of, to nullify any connection the conquered noble had with the supernatural world (Johnston 2001: 380).

Having said all this, we do allow for the possibility that the Nicoya-Guanacaste mirror back *could* represent the result of trade in such "spoils of war." The incised glyphs, as demonstrated above, likely represent an artifact produced by Maya artisans to be gifted between Maya elites. This suggests, at first glance, that its presence in lower Central America might be the result of somewhat suspicious trade activity. The entire argument might be more easily resolved if there were better provenience and artifact associations for the mirror itself. However, we hold out the possibility that this Maya mirror, despite its obvious authenticity and ties with real Maya nobility, may have arrived in lower Central America through other, noneconomic avenues. We cannot rule out that it may have also been a highly-valued, gifted heirloom—an exchange concept that is discussed in greater detail below.

Finally, while the possibility exists that at least one of these lower Central American specimens may have arrived there as the result of economic trade, we do not believe that this provides a blanket explanation for every example. If all the mirrors were traded off following the desecration of Maya royal tombs by rival Maya groups, we would expect the inscriptions and/or iconographic representations on *all* the mirrors to conform to traditional Maya epigraphic organizational canons—as is typical for inscribed mirrors belonging to individual members of the elite in the Maya region (Blainey 2007). The fact that the carved glyphs on the northeast Honduras and La Fortuna mirror backs, for example, do not follow traditional Maya canons suggest that they may have either been emulations or intentionally created "incorrectly" with an alternative purpose in mind.

DISCUSSION

Arguably, these traditional interpretation models are overly simplistic and reflect the strong economic emphasis and Maya-centric bias that colored Central American archaeology in the 1960s and 1970s (and even today). That said, lower Central American artifacts are poorly represented in the Maya archaeological record and problems arise in attempting to explain what exactly it was they were exchanged for if these were, indeed, economic trade items. This lacuna has led many researchers along lines of argument based on negative evidence—perishable materials and/or objects that have not been preserved in the archaeological record. Potential perishable trade goods argued to have been desirable or "acceptable" to most Mesoamerican groups include cacao, feathers, hides, slaves, bark cloth, textile dyes, and cotton textiles, among others (Creamer 1992; Sharer 1984; Snarskis 1983). Regardless, we have problems reconciling why Maya nobility would trade an elite ritual and status item for textile dye, animal hides, or even cacao—all of which are available, in some form or from some location, in the Maya subarea. Obviously a deeper exploration of Maya mirrors in lower Central America is not only timely, but necessary.

NEW(ER) APPROACHES TO INTERPRETATION

EMULATION HYPOTHESIS

Although the majority of examples discussed throughout this chapter are understood to be "real" Maya mirrors created by Maya artisans, the carved example from the site of Wankybila in northeast Honduras is not so certainly of Maya origin. We argue here, based largely on the problematic iconographic elements discussed above, that it likely presents a local emulation of Maya elite status objects.

The Merriam-Webster dictionary defines emulation as the "ambition or endeavor to equal or excel others," generally through the act of "imitation." The emulative act, however, does not occur in a vacuum. It has a purpose, though this purpose may vary between individual actors or cultures, when looked at in archaeological terms. Daniel Miller (1985: 185) has suggested that in past archaeological cultures "the actual process of emulation occurs when hierarchy has developed as a fundamental principle of social organization." As discussed above, there were marked developments in sociocultural complexity taking place by AD 500 throughout lower Central America and this is certainly true of northeast Honduras (Dennett 2007). There we see the beginnings of sociopolitical complexity (and inklings of hierarchy) taking hold, and which roughly coincides with the deposition date-range for mirror fragments at the site of Wankybila.

Developing research suggests that the emulation of Mesoamerican status objects is not unknown from the northeast Honduras archaeological region. Dennett, Luke, and Healy (2008) have recently demonstrated the local emulation of another diagnostic elite item—the Ulua-style marble vase—that was produced at the site of Travesía and made of marble derived from a circumscribed local source zone (Luke and Tykot 2007). In northeast Honduras local potters emulated central Honduran (Mesoamerican) elite styles, creating ceramic versions or copies (although many examples show local symbolic influence in the iconographic program) of the Ulua-style vases. Dennett, Luke, and Healy (2008) argue—along the same lines that Begley (1999) has for the emulation of aspects of Mesoamerican cosmology in large-scale public constructions in the region—that this type of emulation was employed as a power-gathering strategy designed to establish a status-class object while simultaneously nullifying external, Mesoamerican ideological influence in the region.

Ultimately, geochemical sourcing of the mirror-back material may be the only way to resolve the question of whether or not the Wankybila fragments are "real" or "knockoffs." Until then, the mirror fragments from northeast

Honduras suggest some degree of connection—real or imagined—or attempts at disconnection between the developing elite of this region and the established elite of the Maya area.

"Peer Elite" Hypothesis

Undoubtedly, there was purely economic commodity *and* prestige good exchange occurring between southern Mesoamerica and external regions beyond the southeast periphery. Many of these exotic "purchases" also likely served as objects of sociopolitical aggrandizement among emerging chiefs of lower Central America during the Classic period. However, we reiterate that this does not serve as an "all-in-one" blanket explanation. Instead, we suggest that many of these mirrors may have arrived at their destination as a result of symbolic sociopolitical exchange, or "gifting," between "peer elites." Indeed, there is evidence that the finest examples of Maya polychrome painted ceramics and rare raw ores were used as "social currency," gifted between Maya elites of allied territories, and possibly to elites of territories beyond Mesoamerica (Blainey 2007: 188). The idea of gifting between peer elites has also been explored between Maya and Toltec groups from Mexico in the Late Classic period (McVicker and Palka 2001), perhaps indicating that the notion of peer gifting was a well-established practice of the ancient Maya elite.

The concept of gifting is a very old idea in the fields of anthropology and sociology and is typically associated with pre-market or non-state societies. Gifting is most frequently discussed in terms of *reciprocity* and the archaeological definition is "exchanges between individuals or communities who are symmetrically placed, that is they are exchanging things more or less as equals: a sort of gift exchange. One gift does not have to be followed by another at once, but an obligation is created every time a gift is given" (Darvill 2002). In essence, these types of exchange are based on social rather than economic relationships (the objects being gifted are *not* commodities) and, as the definition states, are generally carried out by peers, or equals. The movement of ancient Maya mirrors to lower Central America, we suggest, may represent symbolic exchanges carried out between self-perceived (whether real or imaginary) peer elites in order to develop some form of social connection or, perhaps better stated, commitments or obligations to be reciprocated at a later time (Mauss 1966).

In a situation that did not involve the direct extraction of material goods through force, we argue that leaders—be they emerging chiefs or established kings—would likely want to create a relationship of trust and reciprocal responsibility with distant entities they desired to draw into their sphere of interaction. As such, there would have been a need and/or desire for some form of "social contract" to be developed *before* undertaking other types of "business," which would likely become ultimately economic in nature. Almost a century ago, Mauss (1966: 1) identified this process as a culturally universal occurrence, stating that gifts are "in theory voluntary, disinterested and spontaneous, but are in fact obligatory and interested. The form usually taken is that of the gift generously offered; but the accompanying behaviour is formal pretence and social deception, while the transaction itself is based on obligation and economic self-interest."

In approaching this new hypothesis as a potentially valid one, we must address one fundamental question: Are we able to assess whether or not the mirror was a gift? We argue here that the answer might lie in the mirrors themselves. The odd glyphic and/or iconographic arrangement on some of the mirrors discovered to date in lower Central America might have very well been produced in such a way on purpose. If Maya nobles were gifting to peer elites (or at least attempting to create an atmosphere of equality, to whatever end) in lower Central America, we argue that it might be "sensible" that they did not gift *actual* status items but, rather, an audience-oriented representation. In fact, this practice can be seen among other classes of Mesoamerican status artifacts recovered from lower Central America.

Gifting with the Audience in Mind

One particularly poignant example of "gifting with the audience in mind" between the Mesoamerican southeast periphery and lower Central America is witnessed in carved marble vases (ca. AD 600/650–700/750) from the Ulua River valley in Honduras. Within Mesoamerica proper, marble vases of the Ulua-style from this time period demonstrate uniform Maya design styles (e.g., see figure 11.5a). However, a contemporaneous carved marble vase found in an elite burial in Greater Nicoya (in Guanacaste, Costa Rica) was produced in a distinctly non-Mesoamerican decorative fashion (figure 11.5b), despite exhibiting a chemical signature sourcing the piece to the Ulua valley and traditional, diagnostic Ulua-style carving techniques (Luke and Tykot 2007). The Mesoamerican-produced Ulua-style marble vase instead recalls decorative design styles like that seen on pecked stone bowls from northeast Honduras and, perhaps more importantly, polychrome forms from both the Ulua valley and Greater Nicoya. From roughly AD 500–800 in Greater Nicoya, both Carrillo and Galo polychrome types, in many cases, share stunningly similar design characteristics with well-known Ulua Polychrome types, specifically Paloma (AD 600–700) and Bombero (AD 700–800) polychrome styles. Here we are most concerned with the diagnostic cylindrical vase form and, more particularly, the bird-head lugs seen in both Carrillo (figure 11.5c) and Paloma polychromes. Marble vase and ceramic lug styles are quite particular in the Ulua Valley at this time and although this particular bird-head form is diagnostic to the Honduran Paloma type, it is not typical for marble vases. Rather, and perhaps more importantly, it is a specific style with far greater time depth in Costa Rica. The directionality of influence in this case is not yet fully understood,[4] but we argue that the connection here goes beyond simple one-sided emulation. Instead, we believe that it represents shared symbolic content and sociocultural ties, whether real or perceived. With this ceramic evidence in mind, we argue that the Ulua-style marble vase from Greater Nicoya may have been intentionally designed with its audience in mind—an elite gift from an Ulua valley noble to a peer (or perhaps soon to be?) elite in distant Greater Nicoya. Thus, it seems plausible that the same type of peer gifting may have also been occurring with iron-ore mirrors.

Gifting, by definition, is reciprocal in nature. So, what were the elites of lower Central America gifting back? Why do we not see artifacts in elite burial contexts that may have served as reciprocal gifts? One possible explanation is that the reciprocal gifts were not interred with the dead, as we see in lower Central America. Instead, other Maya ritual contexts may hold the clues that shed light on this query.

The Cenote of Sacrifice (a large, water-saturated sinkhole of collapsed limestone) at the Maya site of Chichen Itza is famous for the vast array of sumptuous pre-Hispanic artifacts found within it. Exotic artifacts from Costa Rica and Panama—including greenstone carvings and elaborate gold pendants and figurines—have been discovered in the cenote (Coggins 1984; Saunders 2003: 31). One particular form of tumbaga bell further demonstrates the idea of gifting with the audience in mind. Rare tumbaga monkey pendants from Panama are generally considered to have been elite status items and comparable examples have been found in the Cenote of Sacrifice (Coggins 1984). However, the specimens from the cenote are generally (and perhaps intentionally) lacking the diagnostic and symbolic cosmological "accessories" of many examples found throughout lower Central America. They also focus strongly on the "bell" aspect of the artifact, a style that would have been aesthetically pleasing to Mesoamerican audiences at the time. We suggest that this type of elite status object may well represent one example of what the peer elite of lower Central America were gifting back.

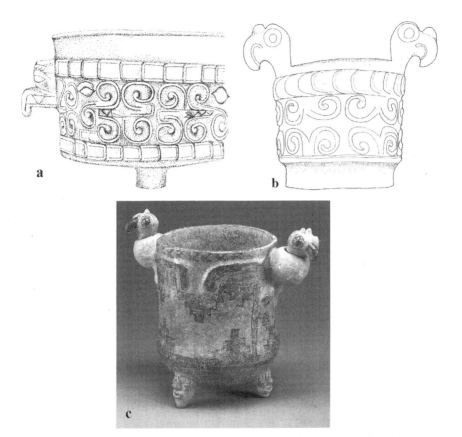

FIGURE 11.5. *(a) Line drawing of a typical Ulua-style marble vase ca. AD 600/650–700/750 (drawing by, and courtesy of, Christina Luke); (b) line drawing of the Ulua-style marble vase from Greater Nicoya (drawing by, and courtesy of, Christina Luke; see Luke and Tykot 2007); (c) Carrillo Polychrome cylinder vase, AD 500–800, Greater Nicoya (Denver Art Museum, Collection of Frederick and Jan Mayer, photography © Denver Art Museum).*

THE RITUAL DISPOSAL OF GIFTS

It seems particularly relevant to note that the alteration or deletion of elements containing cosmological significance appears to have been performed by both Mesoamerican and lower Central American artisans producing materials destined for peer-elite exchange. Perhaps more important than the exchange of these mindfully altered objects, however, is the fact that they appear to have been "disposed of," or taken out of the public domain, in a unique fashion in each cultural realm. John Hoopes (2008) has suggested that Mesoamerican groups may have devalued objects and materials they were unable to acquire or produce in their own realm as a means of nullifying socioeconomic and/or sociopolitical competition. Gifts from peer elites may have been highly valued for the relationship they represented rather than the prestigious nature of the item or its cosmological significance. Indeed, the ritual disposal of gifts in both cultural realms—mortuary contexts in lower Central America and "sacrifice" contents in the Maya area—may have served several complementary ends: an offering to the gods, a show of respect to the gifter, and the removal of competing material culture and ideological influence from external social realms.

Perhaps this explains why so many exotic prestige items from lower Central America are found "disposed of" in ritual places such as the Cenote of Sacrifice in Chichen Itza. Yet, parallels seem to also exist in lower Central America. For example, the fact that a Maya mirror from the site of La Fortuna was deconstructed and reworked to form a necklace of multiple fragment "plaques" (see Stone and Balser 1965) may indicate ritual disposal through its reconfiguration. This aspect of disposal, the reworking/reconfiguration of Maya status objects (often before interment in mortuary contexts), is not unique to iron-ore mirrors—examples of reworked status-related Maya jade celts are also known from Costa Rica after AD 500 (Saunders 2003: 31). Finally, further evidence of the ritual disposal of gifts may be witnessed in the inscribed example from La Fortuna (see figure 11.4) that has three cut marks, which penetrate completely through the stone mirror back. While it is possible that this represents the arrested development of a deconstruction program, they may also represent "kill" marks designed to deplete any Mesoamerican cosmological or sociopolitical significance.

DISCUSSION AND CONCLUSIONS

Following the expansion of Chibchan–speaking, lower Central American groups into regions abutting the Mesoamerican southeast periphery in the early Classic period (ca. AD 300), we see increasing contact and exchange of both commodities and elite status items between the two culture areas across time. The sheer proximity of these groups may have initially influenced the Maya to actively establish a relationship with their more southerly neighbors, perhaps one based on both economic interests and mutual respect. Diachronic shifts in the Mesoamerican power base—from the Maya Southern Lowlands in the Classic period to the Maya Northern Lowlands in the Postclassic period—parallel a shift in the styles of "gifts." In the Classic period shiny *stone* objects were tied to shamanism and elite status. The shift into the Postclassic period, however, sees shiny *metal* objects replace but continue to retain similar (if not the same) religious and sociopolitical connotations (Saunders 2003: 33), perhaps signaling a continued acknowledgment of lower Central American elites as "peer players" on the pre-Hispanic landscape until the time of European Conquest.

In this chapter we have discussed Maya iron-ore mirrors and their occurrence at sites in regions of lower Central America, with a particular focus on Costa Rican sites in the Atlantic Watershed and the Nicoya-Guanacaste subregion of Greater Nicoya. While mirror fragments from northeast Honduras appear to have been local emulations, the specimens from Costa Rica are likely examples of "real" Maya mirrors excavated from contexts outside the Maya region. Geochemical sourcing of all the mirrors discussed throughout this chapter would be most welcome and would certainly go a long way in supporting or refuting some of the arguments presented herein.

Finally, we have argued for more complex interpretive modeling in the explanation of Maya mirrors below the Mesoamerican southeast periphery. While not rejecting traditional models based around hypotheses of chiefly aggrandizement and economic exchange, we have introduced here complementary hypotheses revolving around concepts of emulation and developing peer-elite relationships that we believe more accurately and fully characterize the range of mirrors found in lower Central America.[5] Aspects of these arguments are still being constructed and require further refinement, but we believe that this chapter presents researchers with new questions to address and hypotheses to test. In this respect, we hope that this chapter provides a stronger starting point for future research into the presence of Maya mirrors in lower Central America.

Acknowledgments: This chapter is dedicated to our mentor, Paul F. Healy of Trent University, Ontario, Canada. The original ideas contained in this chapter were first presented in 2008 at the 41st Annual Meeting of the Canadian Archaeological Association held at Peterborough, Ontario. Any errors or omissions are our own.

NOTES

1. Lower Central America is alternatively known as "the Intermediate Area," "the Isthmo-Colombian Area," and "southern Central America."

2. Samuel Lothrop (1937) reported mirror backs from the site of Coclé in central Panama. He describes pyrite mosaic mirrors that do not *seem* Mesoamerican

in character (e.g., holes drilled in the center of the backing and cast-metal frames). Unfortunately, nothing has been published on this since, so they are not discussed here.

3. See also Cuddy (2007; figure 6.7) for a reproduced drawing from Strong's original field notes. Strong's reconstruction contains iconographic elements that were not *at all* apparent to Dennett when analyzing the actual fragment in 2006.

4. The connection between the specific Honduran and Greater Nicoyan ceramic styles discussed here is currently being examined by Rosemary Joyce (personal communication to Dennett 2012). We hope that her research will better inform us with regard to the form of sociocultural relationships examined herein.

5. Maya mirrors have also been recovered in the American Southwest dating to post–AD 550, beginning with Early Hohokam culture (Doyel 2001: 280). As discussed here, traditional interpretations suggest the occurrence of Maya mirrors there are the result of economic trade, but we believe these also deserve reassessment along more sociopolitical lines.

REFERENCES

Baudez, Claude F., and Michael D. Coe. 1966. "Incised Slate Disks from the Atlantic Watershed of Costa Rica: A Commentary." *American Antiquity* 31 (3): 441443. http://dx.doi.org/10.2307/2694760.

Begley, Christopher T. 1999. "Elite Power Strategies and External Connections in Ancient Eastern Honduras." PhD diss., University of Chicago.

Blainey, Marc. 2007. "Surfaces and Beyond: The Political, Ideological, and Economic Significance of Ancient Maya Iron-Ore Mirrors." Master's thesis, Trent University, Peterborough, ON.

Coggins, Clemency Chase. 1984. "The Cenote of Sacrifice: Catalogue." In *Cenote of Sacrifice: Maya Treasures from the Sacred Well at Chichen Itza*, ed. Clemency Chase Coggins and Orrin C. Shane, 23–166. Austin: University of Texas Press.

Cooke, Richard. 2005. "Prehistory of Native Americans on the Central American Land Bridge: Colonization, Dispersal, and Divergence." *Journal of Archaeological Research* 13 (2): 129–187. http://dx.doi.org/10.1007/s10804-005-2486-4.

Creamer, Winifred. 1992. "Regional Exchange along the Pacific Coast of Costa Rica during the Late Polychrome Period, A.C. 1200–1550." *Journal of Field Archaeology* 19: 1–16.

Darvill, Timothy. 2002. *The Concise Oxford Dictionary of Archaeology*. Oxford: Oxford University Press.

Dennett, Carrie L. 2007. "The Rio Claro Site (A.D. 1000–1530), Northeast Honduras: A Ceramic Classification and Examination of External Connections." Master's thesis, Trent University, Peterborough, ON.

Dennett, Carrie L., Christina Luke, and Paul F. Healy. 2008. "Which Came First? The Marble or the Clay? Ulua-Style Vase Production and Pre-Columbian Cultural Boundaries." Paper presented at the 73rd Annual Meeting of the Society for American Archaeology, Vancouver, British Columbia.

Doyel, David. 2001. "Late Hohokam." In *Encyclopedia of Prehistory*, edited by Peter N. Peregrine and Melvin Ember, vol. 6, North America, ed. Peter N. Peregrine and Melvin Ember, 278–286. New York: Plenum Publishers.

Healy, Paul F. 1993. "Northeastern Honduras." In *Pottery of Prehistoric Honduras: Regional Classification and Analysis*, edited by John S. Henderson and Marilyn Beaudry-Corbett, 194–213. Institute of Archaeology Monograph No. 35. Los Angeles: University of California-Los Angeles (UCLA).

Healy, Paul F., and Marc G. Blainey. 2011. "Ancient Maya Mosaic Mirrors: Function, Symbolism, and Meaning." *Ancient Mesoamerica* 22 (02): 229–244. http://dx.doi.org /10.1017/S0956536111000241.

Healy, Paul F., and Carrie L. Dennett. 2006. "The Archaeology of Northeast Honduras and the Isthmo-Colombian Area." Paper presented at the 52nd International Congress of Americanists, Seville, Spain.

Helms, Mary. 1979. *Ancient Panama: Chiefs in Search of Power*. Austin: University of Texas Press.

Helms, Mary. 1992. "Thoughts on Public Symbols and Distant Domains Relevant to the Chiefdoms of Lower Central America." In *Wealth and Hierarchy in the Intermediate Area*, ed. Frederick W. Lange, 207–241. Washington, DC: Dumbarton Oaks.

Hoopes, John W. 2005. "The Emergence of Social Complexity in the Chibchan World of Southern Central America and Northern Colombia, AD 300–600." *Journal of Archaeological Research* 13 (1): 1–47. http://dx.doi.org/10.1007/s10814-005-0809-4.

Hoopes, John W. 2008. "Archaeology without Borders (Discussant)." Paper presented at the 73rd Annual Meeting of the Society for American Archaeology, Vancouver, British Columbia.

Hoopes, John W., and Oscar M. Fonseca Z. 2003. "Gold Work and Chibchan Identity: Endogenous Change and Diffuse Unity in the Isthmo-Colombian Area." In *Gold and Power in Ancient Costa Rica, Panama, and Colombia*, edited by Jeffrey Quilter and John W. Hoopes, 49–90. Washington, DC: Dumbarton Oaks.

Houston, Stephen, David Stuart, and Karl Taube. 2006. *The Memory of Bones: Body, Being, and Experience among the Classic Maya*. Austin: University of Texas Press.

Johnston, Kevin J. 2001. "Broken Fingers: Classic Maya Scribe Capture and Polity Consolidation." *Antiquity* 75 (288): 373–381. http://dx.doi.org/10.1017/S0003598X0 0061020.

Jones, Clarence F., and Paul C. Morrison. 1952. "Evolution of the Banana Industry of Costa Rica." *Economic Geography* 28 (1): 1–19. http://dx.doi.org/10.2307/141616.

Kirchhoff, Paul. (Original work published 1952) 1981. "Mesoamerica: Its Geographical Limits, Ethnic Composition, and Cultural Characteristics." In *Ancient Mesoamerica: Selected Readings*, 2nd ed., ed. John A. Graham, 1–10. Palo Alto, CA: Peek Publications.

Lange, Frederick W. 1992. "The Search for Elite Personages and Site Hierarchies in Greater Nicoya." In *Wealth and Hierarchy in the Intermediate Area*, ed. Frederick W. Lange, 109–139. Washington, DC: Dumbarton Oaks.

Lara Pinto, Gloria. 2006. "La Investigación Arqueológica en Honduras: Lecciones Aprendidas para una Futura Proyección." *Revista Pueblos y Fronteras digital* 2: 1–41. http://www.pueblosyfronteras.unam.mx/a06n2/pdfs/n2_misc2.pdf. Accessed April 15, 2008.

Lothrop, Samuel K. 1937. *Coclé: An Archaeological Study of Central Panama: Part I*. 2 Volumes. Peabody Museum of Archaeology and Ethnology, Memoir No. 7. Cambridge, MA: Harvard University.

Luke, Christina, and Robert H. Tykot. 2007. "Celebrating Place through Luxury Craft Production: Travesia and Ulua Style Marble Vases." *Ancient Mesoamerica* 18 (2): 315–328. http://dx.doi.org/10.1017/S095653610700020X.

Martin, Simon, and Nikolai Grube. 2000. *Chronicle of the Maya Kings and Queens: Deciphering the Dynasties of the Ancient Maya*. New York: Thames and Hudson.

Mauss, Marcel. 1966. *The Gift: Forms and Functions of Exchange in Archaic Societies*. Trans. Ian Cunnison, with an Introduction by E. E. Evans-Pritchard. London: Cohen and West.

McVicker, Donald, and Joel L. Palka. 2001. "A Maya Carved Shell Plaque from Tula, Hidalgo, Mexico." *Ancient Mesoamerica* 12 (2): 175–197. http://dx.doi.org/10.1017/S0956536101122054.

Miller, Daniel. 1985. *Artefacts as Categories: A Study of Ceramic Variability in Central India*. Cambridge: Cambridge University Press.

Miller, Mary Ellen. 1999. *Maya Art and Architecture*. London: Thames and Hudson.

Mol, Angus A. A. 2011. "Bringing Interaction into the Higher Spheres: Social distance in the Late Ceramic Age Greater Antilles as Seen through Ethnohistorical Accounts and the Distribution of Social Valuables." In *Communities in Contact: Essays in Archaeology, Ethnohistory and Ethnography of the Amerindian Circum-Caribbean*, ed. Corinne Lisette Hofman and Anne van Duijvenbode, 61–86. Leiden: Sidestone Press.

Mora-Marín, David. 2005. *"The Jade to Gold Shift in Ancient Costa Rica: A World Systems Perspective."* Manuscript on file. NC: Department of Anthropology, University of North Carolina-Chapel Hill.

Reents-Budet, Dorie. 1994. *Painting the Maya Universe: Royal Ceramics of the Classic Period*. Durham, NC: Duke University Press.

Saturno, William. 2006. "The Dawn of Maya Gods and Kings." *National Geographic* 209 (1): 68–77.

Saunders, Nicholas J. 2003. "'Catching the Light': Technologies of Power and Enchantment in Pre-Columbian Goldworking." In *Gold and Power in Ancient Costa Rica, Panama, and Colombia*, ed. Jeffrey Quilter and John W. Hoopes, 15–47. Washington, DC: Dumbarton Oaks.

Schmidt, Peter, Mercedes de la Garza, and Enrique Nalda, eds. 1998. *Maya*. New York: Rizzoli International.

Sharer, Robert J. 1984. "Lower Central America as Seen from Mesoamerica." In *The Archaeology of Lower Central America*, ed. Frederick W. Lange and Doris Z. Stone, 63–84. Albuquerque: University of New Mexico Press.

Snarskis, Michael J. 2003. "From Jade to Gold in Costa Rica: How, Why, and When." In *Gold and Power in Ancient Costa Rica, Panama, and Colombia*, ed. Jeffrey Quilter and John W. Hoopes, 159–204. Washington, DC: Dumbarton Oaks.

Snarskis, Michael J. (Original work published 1978) 1983. *The Archaeology of the Central Atlantic Watershed of Costa Rica*. Ann Arbor: University Microfilms International.

Stone, Doris Z. 1977. *Pre-Columbian Man in Costa Rica*. Cambridge, MA: Peabody Museum Press.

Stone, Doris Z., and Carlos Balser. 1965. "Incised Slate Disks from the Atlantic Watershed of Costa Rica." *American Antiquity* 30 (3): 310–329. http://dx.doi.org /10.2307/278811.

Strong, William Duncan. 1935. *Archaeological Investigations in the Bay Islands, Spanish Honduras*. Smithsonian Miscellaneous Collections No. 92. Washington, DC: Smithsonian Institution.

Willey, Gordon R. 1966. *An Introduction to American Archaeology: North and Middle America*. vol. 1. 2 vols. Englewood Cliffs, NJ: Prentice-Hall.

Woodbury, Robert, and Aubrey Trik, eds. 1953. *The Ruins of Zaculeu, Guatemala*. Richmond, VA: United Fruit Company.

12

The Huichol Indians of the Western Sierra Madre in Mexico (who call themselves *Wixaritari*), use mirrors for different purposes in a variety of ritual contexts, which are often related to reflexive and creative processes. The main expectation of the ritual actions the Wixaritari carry out with mirrors is to exercise their "gift of seeing" (*nierika*). This skill—which they develop through a long and complex ritual initiation—serves both to become a shaman (singular *mara'akame*; plural *mara'akate*)[1] and for the elaboration of their plastic and aesthetic creations, for example the famous and colorful yarn paintings circulating since the 1960s in the international ethnic art market (figure 12.1).[2] As well as referring to the "gift of seeing" (Neurath 2000: 57–78) of the mara'akate, the term *nierika* also designates a series of objects that can be used by Huichol shamans in the same way as the mirror. Thus, the ancient "front shields" (figure 12.2) (Lumholtz 1900: 108–137) and similar small objects of today made of bamboo structures, circular, triangular, or square (figure 12.3), covered with colorful cotton yarns, are the basic form of the nierika. The elementary morphology of the artifacts clearly underlines their first function as an orifice. On these bamboo circles, threads are sometimes strained to form geometric figures of spider webs or star-like shapes. In this form, they evoke traps formerly used by the Huichol to hunt deer—who are also nierikate—and look like the dream catchers of some Indians of North America. These objects are part of many miniature

The Ritual Uses of Mirrors by the Wixaritari (Huichol Indians)

Instruments of Reflexivity in Creative Processes

Olivia Kindl

DOI: 10.5876/9781607324089.c012

FIGURE 12.1. *Yarn painting of José Benítez Sánchez, Untitled, 2002 (40 × 40 cm)* *(photography: Olivia Kindl, Tepic, 2002; personal collection of Olivia Kindl).*

offerings attached to the votive arrows (*'irɨ*). According to Negrín (1986) and Fresán Jiménez (2002: 65), the central hole of the nierikate functions as a "two-way viewer," because it establishes a communication between different levels of the Huichol cosmos, in the first instance, through the visual faculty.

One of the assumptions I will follow in this chapter is that creative processes imply a reflexive step that can be realized as a visual practice through mirrors in specific ritual contexts. Concerning research on Amerindian cultures in general, it is especially in the study of ritual where reflexivity has become a core subject (Severi 2002: 23–40). Modern and indigenous ritual arts are also often reflective of nature. As it is frequently stated, "Western art

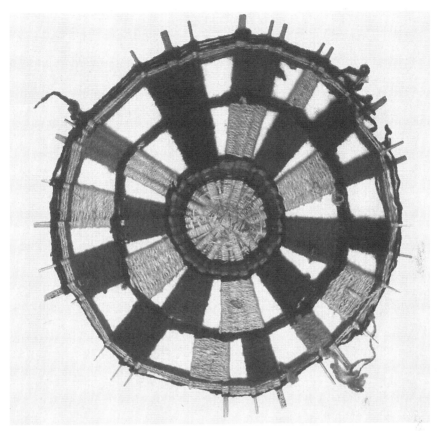

FIGURE 12.2. *"Front shield" with apertures (Lumholtz collection of the American Museum of Natural History, New York, 1894–1897).*

has become a philosophy or criticism of art" (translation by O. Kindl; Belting 2002; Agamben 1993). As can be seen easily in any sale of ethnic arts and/or crafts, indigenous art can also include its own interpretation, ethnographic or iconographic. For example, it may contain the artist's reflection about him/herself, about the artist's society of origin, or the artist's own creative processes. I use the expression about "creative processes" in a broad sense, that is, not only for the development of objects considered as works of art but also for creative gestures that are inserted into ritual and that contribute to its unfolding.

What, where and how does nierika help ritual specialists to see? In response to this question, many of my Huichol informants answered that it is the world of their deified ancestors (*kakaïyari*) that they penetrate thanks to this seeing

FIGURE 12.3. *Offerings on the altar of a xiriki in Las Guayabas, community of Tateikie, San Andrés Cohamiata, Jalisco (photography: Olivia Kindl, 1999).*

capacity. Through these means they are able to learn more about the origin of the diseases of their patients, know what occurs in the case of drought or natural disasters, or discover the true aspect of an evil colleague. However, very little is said about what they actually see.

Consequently, it is necessary to focus not so much on the description of the various objects called nierika—for example, the round-shaped offerings, circular mirrors, circular figures, and face paintings—and what can be perceived through them, but rather on how this specific process of acquiring and manipulating their gift of seeing occurs. This approach requires that we center our interest on the actions realized with these objects, considered as seeing instruments, rather than on their symbolism. In this article, I will focus my analysis on one particular object among *nierikate:* the mirrors. Direct ethnographic observation offers us the privilege to do this. In accordance with Bonhomme's (2007: 2) advice, "the object should not thus be considered *in abstracto*, but always within particular contexts of use, which supposes to privilege the study of these practices above that of their representations" (translation by O. Kindl). The theory I propose is that in the case of the Huichol nierika, the emphasis must be placed on the modalities of vision with respect to the references or contents of these perceptual experiences. This preeminence of the visual process above the fixed image conforms to the reflexions developed by Gombrich (1975: 122) about the distinctions and articulations between visual representations and visual experiences.

So what is the catoptromancy (divination by use of a mirror) for the Wixaritari? How do their shamans use mirrors? What and how can they see through them? Are these ritual uses of mirrors connected with art or creative processes? What is the connection, in Huichol thought, between the mirror, the mask, and the face? Which are the possible associations between the reflective aspect of these mirrors and cognitive-reflexive processes? I will try to connect these questions together by sharing ethnographic fragments that will help us to analyze the phenomena related to the Huichols' ritual use of mirrors.

As we will notice, the uses of the mirror by the Huichols can be observed in two principal types of ritual processes realized through these instruments: divination and initiation. The exercise of both techniques through the specular experiment of perceptions is not restricted to Mesoamerica. Addey (2008: 32–46) mentions similar techniques in her analysis of catoptromancy in ancient Greece, and Bonhomme (2007: 1–16) also points it out for Central Africa and many other parts of the world. A comparative and more universal perspective is thus required to analyze the ritual uses of mirrors. Let us consider three phenomena specific to mirrors: first, the effects of the magnifying glass (change of the proportions); second, the *mise en abyme* (recursive setting in the center); and third, the subject and its double.

THE EFFECT OF THE MAGNIFYING GLASS: MIRRORS AS INSTRUMENTS OF DIVINATION

Most of Huichol mara'akate carry a small circular mirror in a case made from palm leaves called *takwatsi*, where they also arrange their magic instruments, in particular their sticks with attached feathers (*muwieri*) used to purify or cure patients (figure 12.4). As my main Huichol informant explained to me, there are two terms to refer to the mirrors:

> The mara'akame calls its mirror *ne* nierika. *Ne xikiri* is the normal word, for the people who are not mara'akame. Those that are used to see or to paint oneself are *xikiri*. Only those that have nierika, the gift of seeing, can see something in the mirror nierika. Nierika then aids the mara'akate to see or to catch something in their dreams. They see ancestors (kakaɨyari) in their dreams or they listen to them, but they do not see them directly in the mirror. (Translation by O. Kindl; Kɨpaima, Benita Mijares Carrillo, personal communication, 2011).

As I observed several times and as was confirmed by my Huichol interlocutor, the healer uses the mirror to examine his/her patient's body prior to the healing. In order to identify the illness that is causing the pain, the mara'akame applies the mirror to the affected parts of the patient's body and orients it obliquely, in such a way that the mirror reflects the rays of the sun. Some mara'akate have double-sided mirrors, in which case it is not necessary to guide it obliquely, but only to stop it parallel to the body of the patient and see through it exactly like someone using a magnifying glass. The healer looks into the mirror—"as if it were a microscope," Kɨpaima added—and carries out the same process several times, orienting the mirror in different ways and in various parts of the patient's body. This procedure allows the healer to see at the same time the aspect of the disease in its particular form and color (in one case, "it was like a square and very red, it seemed to be inflamed") and to find out the causes of the suffering (e.g., the dissatisfaction of an ancestor for not fulfilling a ceremonial commitment). The feathers of their muwieri are also often used by the shamans to better exercise their "gift of seeing"; among the objects attached to these feather sticks there can also be a tiny mirror, applied on a metal base with the form of a star or sun. Both types of mirrors can be used for the same curing purposes as well as to receive other types of visions sent by the ancestors in particular ceremonial occasions.

Therefore, mirrors have the same function as fires located in the center of the ceremonial enclosures, which the shaman contemplates for entire nights accompanied by his ceremonial chanting. Both mirrors and fires are sources of the visions where the shaman is visited by a deified ancestor, especially

FIGURE 12.4. *Mirror offering and ritual objects of the shaman's case (takwatsi) in the sacred site of Kauyumari Muyewe, Wirikuta, San Luis Potosí, March 2010 (photography: Olivia Kindl, 2010).*

Tamatsi Kauyumari, a cultural hero and the first mara'akame. As a trickster figure (Furst 1997: 97–124), this mythical protagonist plays the role of a clever mediator between the ancestors and humans, and possesses the capacity for switching or moving between different levels of the cosmos, typical of the nierika. These ritual practices are similar to antique Graeco-Roman and Egyptian forms of catoptromancy described by Addey (2008: 40–41), consisting in the use of water and mirrors as conductors of light to invoke beings so that they provide oracular information. The relevance of the light for understanding the visual effects produced by mirrors is explained by Gombrich (1975: 134) as follows:

> The mirror deflects and reflects the light emanating from objects and since we normally assume the light to have come to us along a straight path we believe we see the object behind the mirror. It is indeed tempting to compare a representation with a mirror because both can present a framed surface on which an image appears. But does it really appear on the surface of the mirror? Certainly not if we look with both eyes. Our binocular vision really fuses two different mirror images just as it fuses two different aspects of the three dimensional world in our proximity.

Among the offerings deposited in various holy places in the Huichol territory, one also finds mirrors (see figure 12.4). To offer these to the ancestors is equivalent to providing them with "instruments to see." These ritual objects and other kinds of nierikate permit the entities living on these places to observe the offerings left by the pilgrims and the visual messages they carry. These votive mirrors then also constitute a kind of magnifying glass or, more specifically, a visual conductor by which ancestors can observe living humans. Therefore, in these ritual contexts mirrors constitute kinds of magnifying glasses by which the ancestors and living humans can observe each other.

To the extent that some ceremonial objects are considered as persons—especially the gourd bowls that represent ancestors of the members (Spanish: *jicareros;* Huichol. *xukuri 'ikate*) of the religious organization in a ceremonial center (*tukipa*)—the mirror can also be used to examine these artifacts as if they were the bodies of the ancestors. In this regard, Gutiérrez del Ángel (2010: 84) mentions "a circular mirror called nierika through which, they claim, they can communicate with the deity represented in its ceremonial bowl" (translation by O. Kindl).

As my Huichol interlocutor explained, "before, when there were no mirrors such as those you buy now in stores, there were only a few white stones. They are called *'utatame*, they are white and bright; you can excavate them from the soil, they are a bit opaque and shine" (Kïpaima, Benita Mijares Carrillo, personal communication, 2011). When I asked her if she was talking about the moonstone and showed her some illustrations of it, she confirmed that it was indeed a very similar mineral. She clarified that it was not necessarily used in ritual contexts, but simply to see oneself. She said that with the same purpose, they could use stones of quartz or water (see McGraw, chapter 10, this volume).

In addition, she described to me techniques of lecanomancy, divination practiced with a bowl generally containing water. Similar to the technique of "dream incubation" described by Addey (2008: 32–33) in the case of European antiquity, for the Huichols it is closely related to divination through dreams:

> When the mara'akame baptizes a baby [generally he is the grandfather of the baby], he puts water in a gourd bowl (*jícara*) in the *xiriki* temple. They place it at dusk or at sundown, in order to know what is going to happen, if he can give the baby his name or if another person will give the baby his name, and to know how the baby is going to call. Then he sleeps and dreams the answers. If he dreams promptly, he awakens at midnight and he immediately baptizes the baby; it must be after midnight. When he puts this jícara with water in the xiriki, the baby and his parents must be present. He asks the *teukaritamete* (ancestors) to be with them. Then they prepare themselves for the baptism, and the grandfathers (mara'akate) announce the baby's name and it is baptized. (Translation by O. Kindl, Kïpaima; Benita Mijares Carrillo, personal communication, 2011)

We must not forget the relevance of the shining aspect of the mirror in these divination techniques. According to the explanations of the Wixaritari whom I questioned about this subject, the word for "mirror" in the Huichol language, *xikiri*, refers to things that both shine and are transparent, to items with reflecting surfaces such as those of, for example, crystal, glass, or water. Through association of thought, I was led to the name of a sacred place named *Ha Xikirita*, which means "transparent water" or "water which reflects like a mirror." There is also a female name that derives from the word *Xikirima*, which is translated literally as "transparent" or "brilliant" and which could find its equivalent in the Spanish name Clara. This vernacular definition of the mirror highlights its transparency, its brightness, and the association with water. As in many cases in Mesoamerica, these shining properties of the mirror are particularly appreciated by shamans in exercising their "gift of seeing" (nierika).

In this regard, the manipulation of rock crystals among the Wixaritari should also be mentioned. Called *'irikate* by the Huichols, they appear in the context of ritual procedures and shamanic practices related to healing, ritual hunting of deer, and the cult of the ancestors. The term 'irikame mainly evokes the votive arrow 'iri, in which it is usually attached, in the form of a small stone which is the "last form of life" of a deceased person (Furst and Nahmad 1972: 89).[3] Lumholtz (1900: 63) noted that:

> Small rock crystals, supposed to be produced by the shamans, are thought to be dead or even living people –a kind of astral body of the Theosophists. Such a rock crystal is called *te'vali* (plural, *tevali'r*), or "grandfather," the same name as is given to the majority of the gods. But it may, however, represent any person or relative, in accordance with the directions of the shaman.

The phenomenon of the 'irikame has raised the attention of many researchers interested in understanding the thought and the social organization of the Huichol Indians.[4] Each of them gives explanations from different points of view to understand the phenomenon of the "crystallization of the soul" (Perrin 1996: 403–428) in the form of a small stone of rock crystal. It should be clarified that the 'irikame does not occur in a single form, but it undergoes various changes until it becomes a rock crystal. The shaman catches the 'irikame after having extracted it from the body of the patient with his muwieri. Then he places it in a little cotton ball, wraps it in a piece of cloth and binds it to a votive arrow 'iri. He then sticks it on the roof of the xiriki, where the 'irikame must remain with the ancestors of his lineage. To be able to capture and handle these crystals that are glimmering condensations of ancestors, it is necessary to manage the nierika as the faculty of seeing.

In certain healing sessions, the shaman can see the *kipuri* (a kind of soul) of a person in the form of a transparent and shiny drop of water. This is the counterpart of the living person, which is a "pearl" that remains entrenched in the vault of heaven and dries when the person dies. According to some mythological narratives collected among contemporary Wixaritari, this celestial pearl, which is like a "drop of life," is located at the tip of a huge invisible thread linked to the kipuri of every living person, located at the level of the skull's fontanel.

The definition of the nierika as a "clear vision" and the emphasis on the glossy, transparent or crystalline appearance of things and messages from the gods has implications for aesthetic criteria in Huichol contemporary art (Kindl 2005: 225–248). The yarn paintings, also called nierika, generally exhibit vivid colors, with contrasting tones of light and dark with *moiré* effects; modern materials with iridescent effects, such as acrylic yarns or glass beads, are harnessed by the Huichol artists to produce these visual effects.

The Huichol aesthetic value of luminous materials is closely related with the concept of "shininess" and the "reflective surface complex" analyzed by Healy and Blainey (2011: 229–244) in relation to ancient Maya mosaic mirrors and epigraphy. Olivier (1997: 293–294, 297–298) notes certain techniques of divination with rock crystals, which are still practiced today among some contemporary natives of Mexico. He mentions an ethnographic example in which crystals are illuminated with candles or torches to obtain messages from ancestors, just like the ancient god of the Mexicas Tezcatlipoca did when he made images appear in his sparkling mirror (see also McGraw, chapter 10, this volume).

As Seler (1908 [1901]: 355–391) has shown, there is a link between the ritual use of Huichol nierika and the ancient use of a particular ritual objects among the Mexicas. Called *tlachieloni* or *itlachiaya*, "instruments to see," these artifacts are described by Sahagún (1989: 49, 42) and Durán (1880: 98–99). According to these sources, these ceremonial objects were part of the attributes of two Aztec deities: Xiuhtecuhtli, the God of fire, and Tezcatlipoca, the "Lord of the Smoking Mirror" (Seler 1908: 366–367). On the other hand, it is known that in pre-Hispanic religions of Mesoamerica, mirrors of both obsidian and pyrite had magic uses for communicating with the supernatural world, as well as a symbolism connected with the underworld, metamorphosis, and shamanism. The deployment of meanings, objects, and materials that crystallize around the central figure of the mirror was synthesized by Taube (1992: 198) who, in his detailed analysis of the iconography of the mirror at Teotihuacan, borrows many examples of the Wixaritari, indicating precisely many facets of the nierika:

> It has been noted that Teotihuacan mirrors did not simply symbolize one object but were identified with a wide range of things, such as eyes, faces, flowers, butterflies, hearths, pools of water, webs, woven shields, and caves. At first sight, this may appear strange, but it is clear that among later peoples of Mesoamerica, mirrors were also thought of in a variety of ways. Thus among the modern Huichols, mirrors are considered to be faces, fire, the sun, and caves, and they are linked to a wide variety of other objects having similar circular forms.

This articulation of elements is very close to that found in other contemporary Mesoamerican societies. As evidenced by Olivier's (1997: 288) reference to a Mazatec story, the mirror and water can be replaced by a slit or a hole, which are similar to the "optical tools" used by the ancient Mexicas.

As Preuss (1909: 151) explained, these objects also constitute "tools that gods need to be able to carry out their activities for the good of the world and the human beings." For each category of ritual objects, three levels of significance can be distinguished: prayer, exchange medium, and magic instrument (Kindl and Neurath 2003: 437). As cosmic tools, each category of objects has specific

functions; as we have explained, offering nierikate to the ancestors is equivalent to providing them with "instruments to see." Yet another significant feature of these "optical tools" in the past and the present is their double aspect.
We know that the mirror of Tezcatlipoca also had two sides, to enable men
and gods to "see and be seen" (Olivier 1997: 283–284, 302). This characteristic
also refers to the role of the nierika as a bidirectional viewfinder, allowing
humans to see the face of their ancestors, who in turn can observe humans
through these objects via offerings that are dedicated to them (Fresán Jiménez
2002: 65; Kindl 2005: 242–243).

The uses of the mirror as a magnifying glass in the cases previously described
demonstrate that it is not a question of a simple vanity tool used to contemplate oneself. The mirror is for the mara'akame one of the devices to refine his
vision so that he can receive the messages from the ancestors or inhabitants
of the "world-other" (Perrin 1994: 195–206). At this point, it is important to
understand how the cosmos is structured in the conception of the Huichol
Indians, as it is described in their mythical narrations, and as it appears in their
ritual configurations of space.

THE *MISE EN ABYME*: A MULTICENTERED COSMOGRAPHY

If we explore the network of meanings of which the nierika constitutes the
starting point, we discover that it includes not only the skill to see as well as
various associated ritual objects, but also iconographic figures (peyote,[5] flower,
star, circle, and spiral) and yellow facial paintings (*'uxa*). In addition, it refers
to various elements of the natural and social environment of Huichols and
others inspired by their cosmology. Thus, depending on contexts, nierika may
refer to the sun, peyote cactus (see figure 12.4), the circle of wax decorated with
five accounts of beads located in the center of the gourd bowls, the ceremonial
circular building (*tuki*), offering holes located in the center of the corn field
(*milpa*), the sacred springs, the fire ignited in the center of the ritual spaces, or
even the center of the world, as Wixaritari conceive it. Let us also consider the
bundles of offerings left on the banks of the sacred springs—whose Spanish
name translates literally as "eyes of water" (*ojos de agua*)—that are reflected in
the water (figure 12.5). This optical effect illustrates perfectly the way in which
these sacred springs are also mirrors and passages to the world of the ancestors. The conceptual configuration of the Huichol nierika also includes the
association between mirrors and the earth's surface. This cosmography evokes
the myths of Huichols and Coras describing the shape of the world, where the
Earth's surface is a great platform floating on water.

FIGURE 12.5. *Offerings deposited in the sacred source of Tatei Matinieri on the road to Wirikuta (photograph: Olivia Kindl, 2000).*

As Preuss (1908b: 600–601) teaches us, the nierika objects, these "instruments to see," also constitute representations (*Abbilder*) of the world. This distinction and articulation between a perceptual faculty and a cosmo-geographical design deserves to be explored in more detail. It can be explained through the significance of the verb *nieriya*, "to see" in the Huichol language which, synonymous with "to know" or "to include/understand," implies that it is a matter of "to see the totality" of the world (Kindl and Neurath 2003: 436). However, this aim can be reached only thanks to the capacity of the individuals "to acquire nierika." For that purpose, as we have mentioned, Wixaritari employ various strategies enabling them to channel their vision and to guide their perceptions.

The association between mirrors and the earth's surface is comparable to similar conceptions in other past and present societies of Mesoamerica. Among the ancient Nahuas, the earth's surface, called Anahuatl or Anahuac, could also be depicted in Aztec iconography by the form of a pectoral ornament or an eye (Olivier 1997: 296–297). In contemporary Otomi cosmography, as Galinier (1990: 491) observes, "the underworld is conceived as a specular image of the community." Links between catoptromancy and deities of the water and the earth have also been indicated in the case of ancient Graeco-Roman forms of religion and divination (Addey 2008: 37).

The small central disk of the Huichol's ritual gourd bowls, also called nierika (Kindl 2003), is considered as the earth's surface around which figures emerge: the maize plants grow from there, wax characters and animals walk on this central nierika. On stone disks or votive wooden boards also called nierikate, the center has an equivalent meaning and the hole or central symbol can be replaced by a small circular mirror. Square or round, these small planks of wood—covered with multicolored yarns fitted in a layer of wax—support the drawings that constitute petitions addressed to the sacred ancestors. On the oldest versions of these, the presence of cotton chips—replaced today by acrylic wire—indicates a petition for rain due to the association of cotton with clouds. Today, one finds modern counterparts among the objects the Huichols sell to non-Huichols, regarded in general as a craft industry.

The mirrors appearing in the center of the objects that I have just mentioned belong to a range of artifacts and forms corresponding to a recurring structure in the shape of quincunxes (such as we see on dice in the arrangement of the dots representing the number five). The latter is one of the most prominent common denominators of the ethnic groups of the region of Gran Nayar, including Huichol, Cora, Tepehuan, and Mexicanero Indians. They share the organization of space starting from a fundamental design: the cardinal center and four points. However, this same figure—also called nierika or *tsikiri* in the Huichol language and *chánaka* in Cora (Jáuregui 2003: 251–285)—can simultaneously appear in the center and be repeated around it in a symmetrical and circular arrangement of the symbols.[6] If one examines the place of the nierika-mirror in this composition (as illustrated, in part, by figure 12.6), we observe that it joins together the following characteristics in a recurring way:

Localization of the nierika in the center: concentric insertion of the same
  elementary symbol.
Positive/negative contrast (the mirror, hole, or round figure in the center and
  the objects or the figures around it; this opposition can also be expressed by
  the contrast of the colors).
Combination of several elementary symbols: the nierika can be at the same
  time a peyote, a flower, a star, a circle, a spiral, a hole, or other symbols.
Continuity between the nierika in the center and the figures around it (extension of the symbol in spiral form or geometrical division).
Mirror and symmetrical effects (concentric, non-axial).

FIGURE 12.6. *Votive wood plank (nierika) with five circular mirrors and yarn figures, held by the jicarero of the Sun (Tayau) for an offering in Haramaratsie (San Blas, by Tepic on the Pacific Ocean), ceremonial center of Muku Yuawi, community of Wautia, San Sebastián Teponahuastlán (photograph: Arturo Gutiérrez, 1996).*

This quincunx form reappears on several objects in their flexible and variable plastic form, as well as in the iconographic figures that decorate them, and also in other artistic creations, such as ceremonial dances, music, or songs. Thus, many artistic and visual practices used on several levels by Amerindian tribes of the Gran Nayar seem to be based on a circular design and/or rhomboidal space closely related to a cyclic course of time and on concentric or spheroidal movements.

In addition, the processes of elaborating all these human creations are usually realized according to rotating or spheroidal, concentric, or eccentric movements in a clockwise or levorotatory fashion. If we consider this spatial configuration from an aerial point of view—for example, consulting a map of the Huichol sacred territory (figure 12.7)—we can note that in the horizontal plane, the four cardinal directions and the center take the form of a gigantic *tsikiri* or *ojo de dios* that is repeated in its internal structure and can be extended outward to include more remote sites as well. In the vertical plane—both of its territory and of the world—the nierika reappears in the place through which the *axis mundi* passes, linking three cosmological levels: the terrestrial world of humans, the underworld of the telluric entities, and the celestial region of the solar ancestors and the souls of the mara'akate (figure 12.8). Thus, the nierika is located on a crossroads linking horizontal and vertical world levels as conceived by the Huichols. At the same time that these forms are displayed in this minimal or basic model, they can expand ad infinitum, exactly like reflections endlessly reproducing themselves in two confronted mirrors.

As they expand and become modified, these forms—whether they are objects, iconographic figures, or mental images—refer to a synthetic concept of the world. Starting from this shape of the quincunx, then, we have a concentric adjustment of the interior toward the outside—according to the principle of *homothetic transformation* (Perrin 1994: 198; Galinier 1997: 66)—and also of the outside toward the interior, as when a mirror reflects an image recursively. The Huichol nierika respects this principle: while being spread in multiple variations, it represents at the same time the synthesis of the universe considered in its totality and its essence. As I have described in previous publications (Kindl 2005: 242–243; 2008b: 425–460; 2008a: 33–57; 2010: 65–97), the seeing and to-be-seen dynamic, as it is executed through the nierika-mirror, refers to a particular theory of reflection (Kindl 2008b: 452–454). It is important to emphasize that this conception—which may be defined as a "reflecting complex" (see Blainey, chapter 9, this volume)—should not be confused with the occidental mimesis, which establishes a distance between the reflected object and its image, according to a metaphorical relationship.

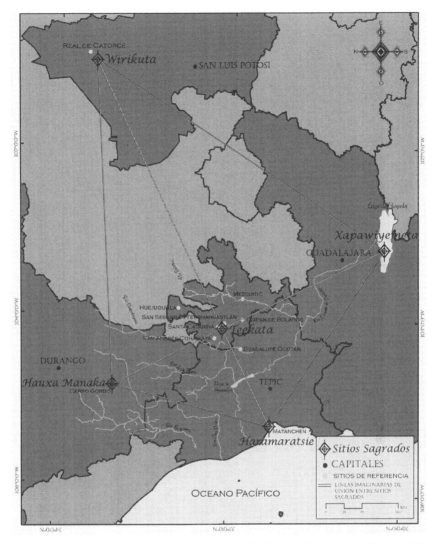

FIGURE 12.7. *Map and geometric overview of the Huichol ritual territory, indicating the distribution of the five major sacred sites for the Wixaritari and the union of such sites from imaginary lines to form a "God's eye" (tsikiri) (map: Bárbara Cristina Lugo Martínez, 2012; sources: INEGI, 2005 and field diaries of Olivia Kindl, 2010).*

In Huichol cosmology, as in many other Mesoamerican societies, the mirror and its equivalents (holes, cuttings, etc.) constitute an interface between different elements, areas, or levels of the world, according to a co-substantial

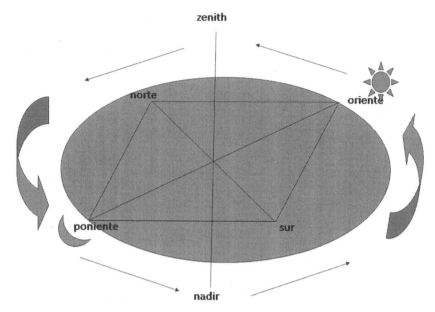

FIGURE 12.8. *Diagram of the Huichol world representation (source: Kindl 2003: 216).*

principle. The exploration of the "Mesoamerican understanding" by Galinier (1999: 101–102; 2009: 153–168) asserts the importance of the "mirror of the world" (Galinier 2009: 158–167) as generator of visions through dynamic processes of metamorphosis, circulation, and transubstantiation. Now that we understand how Huichol cosmography is shaped as a concentric specular space, let us explore the interactions of Huichols with their specular double in initiation processes.

THE SUBJECT AND ITS DOUBLE: A REFLEXIVE INITIATION

In his paper on uses of mirrors in central Africa, Bonhomme (2007: 2) stresses that:

> If the mirror stimulates imaginations at all, it is because it is a strange object. This strangeness comes from the double paradox of specular perception: on the one hand, the reflection of oneself in the mirror duplicates the subject; in addition, spatial reflection is perceived like the prolongation of real space beyond the mirror. The specular double opens onto an identity of giddiness, and specular space onto an ontological giddiness. (Translation by O. Kindl)

The act of the Huichol pilgrim (Spanish:. *peyotero*; Huichol: *hikuritame*) painting his face with a yellow root named *'uxa*[7] by holding a small mirror, is apparently very simple. It may seem to be banal and ordinary, but it implies a complete initiatory preparation. This consists of a gradual growth in mythological knowledge, which is acquired through a series of ritual practices. While this training proceeds in a collective way, it is nevertheless very solitary, because the process is not based on a relation of master with pupil; the shamans give directives, but never explanations. It is up to the initiate, in a dialogue with himself, to deduce the meaning of all that he learns and to understand what he learns. To this extent, the initiation process requires a considerable capacity of reflexivity. Furthermore, by carrying out this simple gesture of painting his face, a series of cognitive operations related to this knowledge will commence.

On several occasions, I observed the application of these ritualized facial paintings, which I consider a creative process. To understand the way a mirror can mediate between a subject and his double in a creative process, let us reflect on what Melchior-Bonnet (1998: 172) detects very precisely in her history of the mirror:

> It is the genius of the artists of the North to have known how to express the
> reflection of the subject on itself, a paradox of an innerness sufficiently "pres
> ent" to make them flourish and distant enough not to impoverish or destroy
> themselves . . . One thinks, for example, of Vermeer's *The Reader* [Girl Reading
> a Letter in an Open Window] . . . ,where a young girl in profile, with an indeci
> pherable face, absorbs herself in the reading of a letter while her leaning image
> is reflected three-quarters in the squares of the window; the screen of a curtain
> underlines the space of the private life; the letter, unreadable to us, the image
> of inaccessible interiority, combines with the reflection, image of concentration
> and reflexivity, to make us feel the impenetrability of the secret, the mystery of
> the one doubling the mystery of the other. (Translation by O. Kindl)

I participated three times in the pilgrimage to Wirikuta[8] with the members of the ceremonial center tukipa of Tunuwametia (located in the center of the community of Tateikie, San Andrés Cohamiata). The cycle of the peyoteros—that is, the pilgrims to Wirikuta—begins right before their departure (in March) and finishes with the peyote festival *Hikuri Neixa* (in May–June). In Wirikuta, several of the pilgrims explained to me that these figures had been inspired by the "ancestors who live over there," and until the dissolution of their group at the end of the dances of the peyote, they would paint these on their faces. As in the case of the use of mirrors—and reminding us that some ceremonial objects have the status of persons—these paintings can also be observed on the paraphernalia that the peyoteros carry with them during their trip, for example the violins of the ritual musicians. These paintings in yellow—the color of the sun and the east—are put on by the peyoteros right before the ingestion of the hallucinogenic cactus. The pilgrims thus carry the mark of the peyote which, by being introduced into their bodies, transforms them into sacred ancestors. The drawings made by painting with the 'uxa refer to these ancestors: for example, the vertical lines evoke rain goddesses like Kaxiwari, the drawings of stags are related to Tamatsi Kauyumari and the other ancestors of the category of Elder Brothers who are associated there. The spirals refer to water gods such as Tatei Nï'ariwame or Tatei 'Utïanaka, which are often represented as snakes rolled up on themselves; the figures of corn seedlings correspond to Tatei Niwetsika and all the other ancestors belonging to "Our Mothers" of corn.

Lastly, the circular figures, which are concentric circles, spirals, or flowers of peyote—whether they are surrounded by rays or not—are identified as the sun and its nierika, or face. Thus, by penetrating the sacred zone of Wirikuta—which, let us recall, is the territory of the rising sun—the face of the peyotero is presented to Tayau or Tawexikïa, Our Father Sun, by his own features, as in a reflective mirror. This to-and-fro of the gaze implies extremely complex plays of vision, which proceed repeatedly from one individual to another and from one point of the cosmos to another.

During the festival of Hikuri Neixa, or the "dance of the peyote," lasting about five days, peyoteros meet to paint their faces. In fact, this process constitutes a ritual in itself, repeated several times during this ceremony, which is celebrated in June or May, at the beginning of the rainy season. In this context, they paint their faces again, with the same figures as during the pilgrimage; this occurs at the beginning of the festival, in the afternoon, often at nightfall. They make the drawings on the apples of their cheeks with the 'uxa, the yellow roots that they collect during their journey from the edge of one of the rare watering points of Wirikuta. Equipped with a flat stone and a small dry grass stem, they rub the roots on the stones by adding a little water to it and then, by soaking the end of the stem in this mixture, they form the drawings on their faces. Some use a small circular mirror for the purpose, those which one purchases from the paint sellers. Some ask one of their companions to paint his face for him and then return the favor. Thus, these face paintings are the result of the visual interaction with oneself or with another self, the *teukari*, or peyotero companion. The drawings produced from this process are of different types and styles according to the individuals, their tasks, and the inspiration of the moment. Once the peyoteros have all painted their faces and their ceremonial objects are also anointed with a little of this yellow pigment, each "carrier of the bowl" (Spanish: *jicarero*; Huichol: *xukuri 'ikame*) associated with the ceremonial center tukipa where the festival will be held, takes the specific paraphernalia in his charge. It is transported in a hand-woven or embroidered bag (Spanish: *morral*; Huichol: *kitsiuri*) filled with the ceremonial objects corresponding to the ancestors he has responsibility to watch over and serve. The group's guide, the 'Irikwekame, has the role of carrying the antlers (including the deer's facial skin, also a nierika in his sense of "face" or "appearance" of an ancestor) as well as the sticks of command (*varas de mando*). These insignia of authority, considered as of solar nature, are also decorated with the yellow paintings of *'uxa*, particularly the deer antlers (figure 12.9). In single file and according to a particular order—corresponding to that respected at the time of the pilgrimages—they move into a nearby thicket. Forming a circle, they deposit all their ceremonial bags and their objects on the ground, which constitute a small mound that acts as an altar. The 'Irikwekame deposits the sticks of command in quite a precise way: they are tilted on the antlers to the west (at the bottom) and to the east (at the top). They then disperse to cut the wood that will burn during the festival in the hearths located inside and outside the circular temple. This task achieved, they sit down on the ground in a circle around the mentioned altar monticule and consume peyote, the hallucinogenic cactus that they brought back from the desert (see also Blainey, chapter 9, this volume). They then bring back bundles of wood to the ceremonial centre and, after having lit the central fire of the large circular building, the peyoteros proceed to the ritual—also repeated several times during the pilgrimage—consisting in the enumeration of the names which they received at Wirikuta. The night that follows is punctuated by sacrifices of animals, circulation around the various places of worship of the ceremonial center, and the ritual known as "the reversal of the fire." At the rising of the sun, the peyoteros wash off their face paintings.

They paint their faces again in the twilight of the following days of Hikuri Neixa. Regulated by circular dances in the court of the ceremonial center tukipa, these days are particularly intense. Around six a.m. the members of the group of the peyoteros separate from the other participants in the festival and will seek their bags containing the ceremonial objects corresponding to their obligations. They move then in the same manner as when they carry out the pilgrimage: in single file, they make the turn of the central hearth inside the circular temple tuki, all blowing in their cow horns or conches, a rallying sign of the peyoteros during the pilgrimage. They arise from the tuki, still in single file, to move to the northwestern side of the ceremonial court. Having

FIGURE 12.9. *'Irikwekame of the ceremonial center of Kuyuaneneme (locality of Las Guayabas, community of Tateikie, San Andrés Cohamiata) with yellow 'uxa face paintings and carrying the deer antlers corresponding to his cargo (photograph: Olivia Kindl, 2002).*

gathered their ceremonial bags and objects, the sticks of command are tilted on the antler to the west-east axis. Then they sit on the ground to paint their faces. There again, either they use small circular mirrors, or their companions draw the figures on their faces. Once the peyoteros have all had their faces painted and their ceremonial objects are also anointed with a little bit of this yellow pigment, they consume peyote again. Then they take their respective bags and ceremonial objects to turn toward the circular building, in the same manner as when arriving, in single file.

At this point in time they proceed to various sacrifices of animals. Generally, they kill cows, bulls, goats, or sheep as offerings for the deified ancestors of the ceremonial centers (e.g., Our Father Sun, Our Grandfather Fire, and Our Mother Maize). As they do for all ceremonial objects and people participating in the ritual, with the sacrificial blood of the animals they anoint the reflecting sides of the *nierika* mirror, be they magical instruments or offerings.

The face painting precedes the ingestion of peyote, closely connecting these two actions. In addition, they also consume peyote (in liquid form) during the days of the dances. As the painting of their faces coincides with nightfall, the figures are erased at dawn. Sometimes the peyoteros complete the face painting just after midday and then wash their faces right after midnight. What then is the connection between transitions from night to day, and darkness to light? According to Huichol cosmology, the sun carries out a daily course around the ground, divided into two principal phases, one downward, the other ascending. From midday on, when it is at its zenith, the sun goes down gradually toward the depths of the underworld, which it traverses during the night. As of midnight its ascending phase starts again.

As a faculty to perceive the world in its completeness, nierika is also closely bound to hearing and the tactile sense, and probably also with the senses of taste and smell. When the peyoteros go to Wirikuta, they penetrate into the world of the ancestors and temporarily lose their identity through ritual practices of inversion, which are realized through language, gestures, the sensory faculties, and from the point of view of the social statuses of members of the group (Myerhoff 1978: 225–295). The efficacy of the nierika to produce ritual inversion processes can be compared to the uses of mirrors described by Bonhomme (2007: 10), in which "through the mediation of the mirror, the initiatory rite contextualizes the specular relation between the ancestors and the living beings" (translation by O. Kindl). In the case of the Huichol pilgrim, what he can see in the mirror is himself (a living person, uninitiated) transforming into another self (an ancestor, initiated).

Thus, the face painting of the peyoteros constitutes one of the distinctive signs that they are initiates. Another of their characteristics is that they sometimes attach these mirrors around their necks,[9] like the Mexican "pectoral," a word that connects this practice to the very ancient objects one observes on many pre-Hispanic statues at the level of the solar plexus. The mirror allows the subject to see his face—this invisible part of the body—and at the same time, through the action of the painting of his face, he can perceive himself as transforming into an ancestor, who is not precisely the peyotero's double, but another person, a being of different ritual and ontological nature. As Bonhomme (2007: 8–9) describes in the case of the initiatory incorporation called *Bwete Misoko* in Central Africa (which also involves the "hallucinogenic" plant called Iboga), to see an ancestor appearing in the mirror is an initiatory process in which the initiate transforms himself and observes his own metamorphosis in the mirror.

The ambivalent operation produced by masks—which simultaneously "dissimulate and gives to see"—also participates in this panoply of "instruments of paradox" (Bonhomme 2007: 11) in the Huichol nierika complex. Indeed, the *'uxa* face paintings can be applied in the same manner to the face as they are used to decorate masks, such as those used by the ritual clowns (*tsikwaki*). These objects are intrinsically related to the concept of nierika, even if they do not bear the name. This is due to the fact that they indicate and deal with the eyes, apples of the cheeks, and—by extension—the whole of the face.

As regards the parallels between the ritual actions carried out by peyoteros during the pilgrimage and those which they repeat during the festival of Hikuri Neixa, one wonders why the ritual actions of the pilgrimage are similar to those of the ceremony: do these ritual actions imitate each other? Do they perform these parallels for themselves, in a self-reflexive way, to make a kind of assessment of their pilgrimage to Wirikuta? Do they do it for the other participants in the festival, those who did not participate in the pilgrimage, as a demonstration? All of these options are possible. In any case, these reflexive processes in ritual actions can also be identified with mirror reflections.

CONCLUSION

Let us return to our starting assumption: among the Wixaritari, creative processes imply a reflexive step in which mirrors—and the concept of nierika related to these—play a central role as instruments of mediation through the visible and the invisible worlds. With regard to Huichols, we can assert that creative qualities initially appear as a capacity not only to preserve and develop a visual perspicacity, making it possible to perceive the surrounding world distinctly, but also to seize the inspiration that the ancestors concede through visionary experiences. This is also confirmed by the making of objects and images in a ritual context, as well as in the field of artistic creation or artisanal production. As we saw, the mirror constitutes a fundamental instrument to acquire this "clear vision." Indeed, clarity and transparency come from the solar world of the celestial ancestors, precisely those which give the Huichols their aesthetic accomplishments but which can also blind them; "a prism such as the mirror can disorganize the field of view, because it hides as much as it shows" (Melchior-Bonnet 1998: 113–114). It is this interstice that makes it possible "to put the world in perspective and to support the reflexive cogito" (Melchior-Bonnet 1998: 114). The passage—hole, opening, or door/portal—cleared by this "instrument of seeing," which is the mirror-nierika, must be maintained in a constant dynamic in order to keep open this liminal and enigmatic interface between the visible and the invisible.

NOTES

1. In some cases, the mara'akame can be compared to a shaman. Healer and connoisseur of the mythology, he is the officiating clergy of traditional ceremonies. He has the ability to communicate with the ancestors through his ceremonial songs, dreams, or visions, with which he can experiment under the effects of psychotropic substances such as peyote.

2. I have pointed out elsewhere (Kindl 2010: 73–74) that this convergence of goals does not necessarily mean that artists have to be shamans.

3. An adult person, who has "already fulfilled," can also deliver his soul to his relatives in the form of a rock crystal.

4. The classical authors Lumholtz (1900: 1–291; 1987), Preuss (1908a: 582–604; 1908b: 369–398), and Zingg (1982) address the issue, as well as Furst and Nahmad (1972), Fikes (1993: 120–148), Negrín (1986), Perrin (1996: 403–428), and Neurath (2002a, 2002b: 96–119, 2004: 93–118), among the most recent.

5. Hallucinogenic cactus called *hikuri* in the Huichol language, botanically classified as *Lophophora williamsii*. Huichols collect peyote in the sacred territory of Wirikuta, the semidesert *altiplano* of the state of San Luis Potosí.

6. For more details concerning the difference between the quincunx division as it mostly appears on ritual objects and the star-like structure in six or more parts as it can be observed on commercial objects, see Kindl (2003).

7. The plant of this root grows in the sacred territory of Wirikuta, where the Huichol congregate and also collect the cactus peyote (*hikuri*). Called *agrito* in the Spanish of the inhabitants of this region, it is botanically classified as *Berberis trifoliata*.

8. Name of a sacred zone located in a semidesert altiplano in the State of San Luis Potosí, more than 400 km away from the Huichol mountains.

9. Small circular mirrors are still worn on the chest as pectorals by the peyoteros and mara'akate of some Huichol communities, especially Tuapurie, Santa Catarina Cuexco-matitlán. According to some interlocutors, they were also used in this way about thirty years ago in Tateikie, San Andrés Cohamiata, but rather for an ornamental purpose.

BIBLIOGRAPHY

Addey, Crystal. 2008. "Mirrors and Divination: Catoptromancy, Oracles and Earth Goddesses in Antiquity." In *The Book of the Mirror: An Interdisciplinary Collection Exploring the Cultural Story of the Mirror*, ed. Miranda Anderson, 32–46. Newcastle: Cambridge Scholars Publishing.

Agamben, Giorgio. (Original work published 1977) 1993. *Stanzas: Word and Phantasm in Western Culture*. Minneapolis: University of Minnesota Press.

Belting, Hans. (Original work published 1995) 2002. *Das Ende der Kunstgesichte. Eine Revisión nach zehn Jahren*. München: Beck.

Bonhomme, Julien. 2007. "Réflexions multiples: Le miroir et ses usages en Afrique centrale." *Images re-vues* 4: document 9. http://imagesrevues.revues.org/147. Accessed February 9, 2012.

Durán, Fray Diego. (1579–1581) 1880. *Historia de las Indias de Nueva España e islas de Tierra Firme*, volume II and Atlas. Mexico: Imprenta de Ignacio Escalante.

Fikes, Jay C. 1993. "To Be or Not To Be: Suicide and Sexuality in Huichol Indian Funeral-Ritual Oratory." In *New Voices in Native American Literary Criticism*, ed. Arnold Krupat, 120–148. Washington, DC: Smithsonian Institution Press.

Fresán Jiménez, Mariana. 2002. *Nierika: Una ventana al mundo de los antepasados*. Mexico: CONACULTA–Fondo Nacional para la Cultura y las Artes.

Furst, Peter T. 1997. "The 'Half-Bad' Kauyumari: Trickster-Culture Hero of the Huichols." *Journal of Latin American Lore* 20–1: 97–124.

Furst, Peter T., and Salomón Nahmad. 1972. *Mitos y arte huicholes*. Colección Sep-Setentas. Mexico: Secretaría de Educación Pública.

Galinier, Jacques. 1990. *La mitad del mundo: Cuerpo y cosmos en los rituales otomíes*. Mexico: Universidad Nacional Autónoma de México–Centro Francés de Estudios Mexicanos y Centroamericanos–Instituto Nacional Indigenista.

Galinier, Jacques. 1997. *La moitié du monde: Le corps et le cosmos dans le rituel des Indiens otomí*. Paris: Presses Universitaires de France.

Galinier, Jacques. 1999. "L'entendement mésoaméricain. Catégories et objets du monde." *L'Homme* 39 (151): 101–121. http://dx.doi.org/10.3406/hom.1999.453621.

Galinier, Jacques. 2009. "Pensar fuera de sí: Espejos identitarios en Mesoamérica." In *El espejo otomí: De la etnografia a la antropologia psicoanalítica*, edited by Jacques

Galinier, 153–168. Mexico: Centro de Estudios Mexicanos y Centroamericanos–Comisión Nacional para El Desarrollo de los Pueblos Indígenas–Instituto Nacional de Antropología e Historia.

Gombrich, Ernst H. 1975. "Mirror and Map: Theories of Pictorial Representation." *Philosophical Transactions of the Royal Society of London. Series B, Biological Sciences* 270 (903): 119–149. http://dx.doi.org/10.1098/rstb.1975.0005.

Gutiérrez del Ángel, Arturo. 2010. *Las danzas del padre sol: Ritualidad y procesos narrativos en un pueblo del Occidente mexicano.* México: Universidad Nacional Autónoma.

Healy, Paul F., and Marc G. Blainey. 2011. "Ancient Maya Mosaic Mirrors: Function, Symbolism, and Meaning." *Ancient Mesoamerica* 22 (2): 229–244. http://dx.doi.org/10.1017/S0956536111000241.

Jáuregui, Jesús. 2003. "El *cha'anaka* de los coras, el *tsikuri* de los huicholes y el *tamoanchan* de los mexicas." In *Flechadores de Estrella:. Nuevas aportaciones a la etnología de coras y huicholes,* ed. Jesús Jáuregui and Johannes Neurath, 251–285. Mexico: Instituto Nacional de Antropología e Historia—Universidad de Guadalajara.

Kindl, Olivia. 2003. *La jícara huichola: Un microcosmos mesoamericano.* México City: Instituto Nacional de Antropología e Historia-Universidad de Guadalajara.

Kindl, Olivia. 2005. "L'art du *nierika* chez les Huichols du Mexique: Un instrument pour voir." In *Les cultures à l'œuvre. Rencontres en art,* edited by Michèle Coquet, Brigitte Derlon and Monique Jeudy-Ballini, 225–248. Paris: Biro éditeur-Éditions de la Maison des sciences de l'homme.

Kindl, Olivia. 2008a. "El arte como construcción de la visión: *Nierika* huichol, interacciones sensibles y dinámicas creativas." *Diario de Campo,* 48 Suplement (May–June): 33–57.

Kindl, Olivia. 2008b. "¿*Imago mundi* o parabola del espejo? Reflexiones acerca del espacio plástico huichol." In *Las vías del noroeste II: Propuesta para una perspectiva sistémica e interdisciplinaria,* ed. Carlo Bonfiglioli, Arturo Gutiérrez, Marie-Areti Hers, and María Eugenia Olavarría, 425–460. Mexico: Universidad Nacional Autónoma de México, Instituto de Investigaciones Antropológicas.

Kindl, Olivia. 2010. "Apuntes sobre las formas ambiguas y su eficacia ritual: Un análisis comparativo desde el punto de vista de los huicholes (*wixaritari*)." In *Las artes del ritual. Nuevas propuestas para la antropología del arte desde el Occidente de México,* edited by Elizabeth Araiza Hernández, 65–97. Zamora: El Colegio de Michoacán.

Kindl, Olivia, and Johannes Neurath. 2003. "El arte *wixarika*, tradición y creatividad." In *Flechadores de estrellas: nuevas aportaciones etnológicas sobre coras y huicholes,* ed. Jesús Jáuregui and Johannes Neurath, 413–53. Mexico: Instituto Nacional de Antropología e Historia–Universidad de Guadalajara.

Lumholtz, Carl. 1900. "Symbolism of the Huichol Indians." *Memoirs of the American Museum of Natural History* 3 (Anthropology): 1–291.

Lumholtz, Carl. (Original work published 1902) 1987. *Unknown Mexico. Explorations in the Sierra Madre and other Regions 1890–1898.* 2 vols. New York: Dover Publications.

Melchior-Bonnet, Sabine. (Original work published 1994) 1998. *Histoire du miroir.* Paris: Hachette Littératures.

Myerhoff, Barbara G. 1978. "Return to Wirikuta: Ritual Reversal and Symbolic Continuity on the Peyote Hunt of the Huichol Indians." In *The Reversible World: Symbolic Inversion in Art and Society*, ed. Barbara Babcock, 225–295. Ithaca: Cornell University.

Negrín, Juan. 1986. *Nierica: Espejo entre dos mundos. Arte contemporáneo huichol.* México: Museo de Arte Moderno.

Neurath, Johannes. 2000. "El 'don de ver': El proceso de iniciación y sus implicaciones para la cosmovisión huichola." *Desacatos* 5:57–78.

Neurath, Johannes. 2002a. *Las fiestas de la Casa Grande: Procesos rituales, cosmovisión y estructura social en una comunidad huichola.* Mexico: Instituto Nacional de Antropología e Historia–Universidad de Guadalajara.

Neurath, Johannes. 2002b. ""Mitos cosmogónicos, grupos rituales e iniciación: Hacia una etnología comparada del Gran Nayar y del Sudoeste de Estados Unidos." *Boletín Oficial del Instituto Nacional de Antropología e Historia.*" *Antropología* 68: 96–119.

Neurath, Johannes. 2004. "El doble personaje del planeta Venus en las religiones indígenas del Gran Nayar: Mitología, ritual agrícola y sacrificio." *Journal de la Société des Americanistes* 90–1: 93–118.

Olivier, Guilhem. 1997. *Moqueries et métamorphoses d'un dieu aztèque: Tezcatlipoca, le "Seigneur au miroir fumant".* Paris: Institut d'Ethnologie.

Perrin, Michel. 1994. "Notes d'ethnographie huichol: La notion de *ma'ive* et la nosologie." *Journal de la Société des Americanistes* 80:195–206.

Perrin, Michel. 1996. "The Urukame, A Crystallization of the Soul: Death and Memory." In *People of the Peyote: Huichol Indian History, Religion and Survival*, ed. Stacy B. Schaefer and Peter T. Furst, 403–428. Albuquerque: University of New Mexico Press.

Preuss, Konrad Th. 1908a. "Die religiösen Gesänge und Mythen einiger Stämme der mexikanischen Sierra Madre." *Archiv für Religionswissenschaft* 11 (2–3): 369–398.

Preuss, Konrad Th. 1908b. "Ethnographische Ergebnisse einer Reise in die mexikanische Sierra Madre." *Zeitschrift fur Ethnologie* 40:582–604.

Preuss, Konrad Th. 1909. "Ethnographische Sammlung aus Mexiko." *Beiblatt zum Jahrbuch der Königlichen Preussischen Kunstsammlungen* 30:150–5.

Sahagún, Fr. Bernardino de. (1547–1577) 1989. *Historia General de las cosas de Nueva España*, 2 volumes, edited by Alfredo López-Austin and Josefina García Quintana. Mexico: Alianza Editorial Mexicana.

Seler, Eduard. (1901) 1908. "Die Huichol-Indianer des Staates Jalisco." *Gesammelte Abhandlungen zur Amerikanischen Sprach- und Alterthumskunde* 3: 355–391.

Severi, Carlo. 2002. "Memory, Reflexivity and Belief: Reflexions on the Ritual Use of Language." *Social Anthropology* 10 (1): 23–40. http://dx.doi.org/10.1111/j.1469-8676 .2002.tb00044.x.

Taube, Karl A. 1992. "The Iconography of Mirrors at Teotihuacan." In *Art, Ideology and the City of Teotihuacan*, ed. Janet C. Berlo, 169–204. Washington, DC: Dumbarton Oaks, Trustee for Harvard University.

Zingg, Robert M. (Original work published 1938) 1982. *Los huicholes: Una tribu de artistas*. 2 vols. Mexico: Instituto Naciona Indigenista.

13

Largely as the result of a session at the Society for American Archaeology (SAA) annual meeting held in Sacramento in 2011, this collection of essays is exceptionally broad in scope as well as in depth in terms of highly focused research, and constitutes the most important and illuminating contribution to our understanding of ancient Mesoamerican mirrors to date. It also complements and builds on other recent valuable studies concerning ancient Mesoamerican mirrors, including works by Marc Blainey (2007), Paul Healy and Marc Blainey (2011), Guilhem Olivier (2003: 240–268), and Tomás Villa Córdova (2010). In the present volume, the first two chapters—one by Emiliano Gallaga and the second coauthored by Emiliano Melgar, Emiliano Gallaga, and Reyna Solis—are devoted to modern replicative lithic studies, a form of experimental archaeology concerning the technology and effort required for the creation of pyrite mirrors. In addition, the contribution by Brigitte Kovacevich discusses ancient pyrite working at the site of Cancuén, Guatemala. Other chapters address archaeologically documented finds of mirrors at such sites and regions as Teotihuacan, Jalisco, Zacatecas, and lower Central America. In addition, the chapter by José Lunazzi analyzes and describes the optic qualities of polished stone mirrors, a topic also very much addressed by other authors in the volume, and discussed further below. Many contributions address the symbolism of ancient Mesoamerican mirrors, including those by Brigitte

Through a Glass, Brightly

Recent Investigations Concerning Mirrors and Scrying in Ancient and Contemporary Mesoamerica

Karl Taube

DOI: 10.5876/9781607324089.c013

Kovacevich, Joseph Mountjoy, and Marc Blainey. It is also important to note the contemporary ethnographic work by Olivia Kindl and John McGraw concerning the divinatory use of mirrors and crystals among the Huichol of Jalisco and the highland Maya peoples of Guatemala. Such anthropological research truly offers windows of access to the ancestral past via contemporary native practitioners.

In this concluding discussion, I address the broadly varied contributions of this volume under a number of themes, these being the ontology of mirror stones both in terms of their working and their natural properties of reflecting light, including contrasts of luminosity that appear to be often diametrically opposed as brightness and darkness in Mesoamerican thought. There is also the symbolism of mirrors as well as the ritual practice of scrying in traditional Mesoamerica, and among the topics to be discussed in terms of this collection of studies is the use of mirrors to see one's own face as a "self-reflective device" for personal introspection. In addition, as I have mentioned in previous work (Taube 1992a, 2001a, 2001b), round mirrors have a broad array of overlapping and by no means contradictory meanings in Mesoamerican thought, including their relation to eyes, faces, flowers, caves and portals, the sun and the world—themes also addressed in the chapter by Olivia Kindl concerning the contemporary Huichol.

The experimental lithic work discussed in chapters 2 and 3 are of critical importance not only for understanding the technology of manufacturing pyrite mirrors, but also the sheer effort that is required to construct these fine objects. Although there has been a great deal of replicative study concerning the ancient manufacture of Mesoamerican flint and obsidian artifacts, there has been surprisingly little research on the working of iron ores, with one noteworthy exception being John Carlson's work creating an Olmec-style parabolic mirror from magnetite (see Carlson 1981: 122–123, figure 3). The experimental archaeology efforts witnessed in Emiliano Gallaga's creation of a circular sandstone backing or base and single pyrite plaque are truly illuminating in terms of the amount of work that went into creating a single mirror, which he estimates at between 100 and 150 days of human labor. As with jade carving and jewelry today, the value of these ancient creations is not only the intrinsic worth of the stone but also the time, skill, and effort invested into fashioning the piece. In these terms, pyrite mosaic mirrors were obviously of great importance in ancient Mesoamerica, although it should be noted that "composite" mirrors composed of minute pyrite fragments glued onto clayey backing and then polished have recently been documented for ancient Mesoamerica, including at the Lake Amatitlan site of Los Mexicanos, which is well known for its ceramic offerings in strong Teotihuacan style (Nelson et al. 2009). As the authors note, such mirrors would be far easier to produce than ones created from cut and tightly fitted pyrite tesserae.

Whether of glued and polished pyrite fragments or solid and carefully carved slabs, both types of pyrite mirror must have been of great value, which is all the more striking when one considers the number of mirrors that have been discovered, including at such a seemingly minor site as Nebaj, in the Alta Verapaz of highland Guatemala (see Smith and Kidder 1951); this brings up questions as to how these precious items were circulated to such distant regions as the Hohokam region of southern Arizona and the Cocle area of Panama (see Gladwin et al. 1938, Kidder et al. 1946: 126, 131–132). In chapter 11, Carrie Dennett and Marc Blainey bring up the intriguing topic as to how such fine and rare objects were exchanged to the Intermediate Area of Central America. I would suggest that the Classic Maya were surely keen on obtaining the tail feathers of the male quetzal, which might well have been overhunted in the Maya region during the Classic period (AD 250–900).

The coauthored contribution in chapter 3 concerns the experimental archaeology of the specific technology used to fashion pyrite and related iron ores into polished mirror surfaces. This research includes the use of microscopic imagery to determine the methods and efforts devoted to constructing ancient Mesoamerican mosaic mirrors, which typically have a base of slate or sandstone with a reflective surface of finely cut pyrite slabs. Clearly enough, the pyrite pieces would have been cut by string sawing, that is, a piece of cordage with a grit harder than pyrite ore, which is about 6 to 6.5 on the Mohs scale of hardness. What the cutting materials were is not clear, and although it remains to be documented in the archaeological record, one possibility along with quartz and jadeite grit is garnet, with a Mohs of roughly 7. Jadeite— an especially prized and coveted stone of ancient Mesoamerica—forms in a stone known as eclogite, which contains garnet crystals (Taube et al. 2004: 213). The cutting and polishing of pyrite surely relates to the ancient technology of jade working from the Motagua region of eastern Guatemala, and for this reason, it is entirely possible that garnet grit and powder were employed as cutting and polishing agents.

The experimental archaeology discussed in chapters 2 and 3 is nicely bridged in chapter 4, with the discussion by Brigitte Kovacevich concerning ancient lithic workshops at Cancuén, Guatemala. In this groundbreaking research, Kovacevich notes the presence of ceramic string saw "anchors" at non-elite households. Not only does this imply that thin pyrite faces were cut by this method, but also that the lithic industry of preparing pyrite mosaic mirrors was probably performed by highly trained commoners rather that an elite palace craft. I do consider this in sharp contrast to much Classic Maya jade working, which is often filled with esoteric elite references to the maize god and other deities.

In terms of experimental archaeology and the nature of mirror lithics, one topic worth addressing is that as an iron ore, pyrite is indeed a "hot" stone, entirely aside from its abilities to attract heat and provide a shiny, lustrous surface. Indeed, the term *pyrite* is from the Greek term for fire, and surely relates to its quality of delivering hot sparks when struck by a hard stone, such as flint. Any ancient Mesoamerican knapper would certainly have noticed this highly unusual trait. That is, aside from being a shining yellow "sun stone," pyrite also provided fire and heat in ancient Mesoamerican thought. As far as I am aware, pyrite was never used as a fire maker in ancient Mesoamerica, although the *Florentine Codex* does mention the creation of sparks and fire from flint or *tecpatl* among the Aztec, despite the fact that it is not possible to create sparks by striking two flints together (Sahagún 1950–1982, bk. 11: 229; Stapert and Johanssen 1999). However, the use of pyrite for firemaking, including striking two pyrite pieces together or with a flint hammer or scraper is documented among the Eskimo and as early as the Mesolithic and probably the Paleolithic periods of northern Europe (Stapert and Johanssen 1999).

Although the chapters of this volume focus on mirrors of pyrite and other iron ores, other materials were also used as reflective devices, including surfaces of water and obsidian, which brings up the topic of luminosity. In this regard, José Lunazzi's chapter concerning the optic qualities of pyrite mirrors is a most welcome contribution. Lunazzi brings up the very important aspect of the optic qualities of polished pyrite mirrors, which relates directly to the experimental archaeology replicating the creation of these objects. As is abundantly clear in this volume, most archaeologically documented pyrite mirrors of ancient Mesoamerica are reduced to an entirely dull, oxidized sulfide powder atop the backing of slate or sandstone. The exceptions are noteworthy, including a large circular mirror discovered in a tomb at Bonampak, Chiapas (see Schmidt 1998: no. 203). From a different part of the New World, however, ancient pyrite mirrors from the Wari and other cultures of the Central Andes could offer vital clues for comparison, thanks to the extremely dry conditions of coastal Peru. For instance, a very well-preserved Wari wooden mosaic mirror at Dumbarton Oaks would be an excellent avenue of approach concerning the optic qualities of ancient pyrite mirrors from Mesoamerica (see Boone 1996: 181–86). In addition, a well-preserved pyrite mosaic mirror wrapped in cotton and deer skin was discovered in a cave near Tempe, Arizona, a piece that clearly derives from ancient Mesoamerica (see Gladwin et al. 1938: plate CXII). Aside from pyrite mosaic mirrors, there is also a solid, convex pyrite mirror portraying on its back surface the Late Postclassic wind god, Ehecatl-Quetzalcoatl, in the Musée du Quai Branly in Paris (Easby and Scott 1970: no. 303).

Although I have mentioned several examples of well preserved, ancient pyrite mirrors, the work by Emiliano Gallaga and others in this volume opens up the concept of not only recreating ancient Mesoamerican mirrors using what is essentially a Neolithic stone technology—obviously a very time-exacting prospect—but also the opportunity to observe the reflective qualities of polished pyrite, that is, studying the optic qualities of polished pyrite and related iron ores under various light conditions; this includes the refractive nature of highly polished iron ore in terms of plays of light, such as at night with a full moon versus the sun at high noon. As someone experienced at cutting and polishing jadeite with contemporary lapidary machines, I am sure that the cutting and polishing of a slab of iron pyrite or grinding and polishing of a circular "marcacite dollar" could readily replicate the optic qualities of polished pyrite stone with relatively little effort. In addition, how would a polished pyrite mirror project light to other surfaces, such as the entry to a dark temple room or cave? According to the ancient Greek philosopher Iamblichus, one form of divinatory scrying, or "catoptromancy," was the projection of light from a mirror onto a wall (Addey 2007: 41). Clearly enough, the creation of polished surfaces of pyrite even with modern lapidary methods could have major implications for understanding the ritual use of pyrite mirrors in ancient Mesoamerica.

Aside from pyrite and other related iron ores, obsidian was also a major mirror stone in ancient Mesoamerica, although such mirrors are oddly absent during the Classic period, including both at Teotihuacan and in the Maya region. There is the celebrated Central Mexican god Tezcatlipoca, or "Smoking Mirror," during the Late Postclassic period, and highly polished, contact-period obsidian mirrors continued to be valued in Mexico into at least the mid-sixteenth century. The *Codex Kingsborough*, a native legal document dating to 1554, describes in great detail the confiscation of precious objects, including gold jewelry and featherwork from the community of Tepetloztcoc, including no less than 10 obsidian mirrors in circular frames. Why the local Spanish authorities desired these pieces remains obscure indeed, but these objects were probably used for ritual purposes by Europeans in the Viceroyalty as well as in Europe. The Doctor John Dee, appointed conjurer to Queen Elizabeth, owned such a mirror, which is currently on view at the British

Museum (Tait 1967; Woolley 2001: 155). There is also a large obsidian mirror in a round gilded wood frame displaying 14 four-petalled blossoms at the Museum of Natural History in New York (see Nicholson and Berger 1968: figure 22). In addition, a rectangular obsidian mirror framed with a wooden backing at Dumbarton Oaks bears the Franciscan stigmata of Christ on the opposite side (see Evans 2010: 76–79).

While obsidian mirrors were clearly widespread in contact-period Mexico, they are also known for Protoclassic West Mexico, that is, roughly two thousand years ago. In this case, a single blow creates the mirror surface by direct percussion, with no grinding and polishing to fashion the surface. One such mirror was excavated in a shaft tomb at Tabachines, on the outskirts of Guadalajara, Jalisco (Schöndube B. and Galvan V. 1978: 154, fig. 23.9). Another similar obsidian mirror attributed to Jalisco is in the collection of the Los Angeles Museum of Natural History (see Mountjoy 1998: figure 11). These early West Mexican mirrors immediately recall the large obsidian disks created by direct percussion that were inserted into the eyes of the great feathered serpent heads and the "War Serpent" helmet masks on the Temple of Quetzalcoatl at Teotihuacan. Dating to roughly AD 250, this structure demonstrates that by at least this time, mirrors were related to eyes in Mesoamerican thought, both as objects of sight but also as gleaming entities that project light and therefore vision as well (for concepts of projective vision, see Houston and Taube 2000). Centuries later, obsidian disks were also employed in finely carved piers from the Palace of Quetzalpapalotl at Teotihuacan, which appear in the eyes of birds as well as rows of goggled human eyes (see Acosta 1964: lám 6, figures 42, 44). This brings up the fact that iron pyrite was also commonly placed in the eyes of Teotihuacan stone masks and statuettes, clearly denoting a contrast between the two mirror stones.

For the Classic Maya, there was a sharp distinction between two types of god eyes, one being that of the diurnal sun deity, which also serves as the basic sign for shiny reflective surfaces, including polished celts and mirrors (figure 13.1a; see Blainey 2007; Schele and Miller 1986). The other eye, however, is spiral and is found with the night sun, or the Jaguar God of the Underworld (figure 13.1b; see Houston and Taube 2000: 283–84). In addition, there are Classic Maya portrayals of mirrors with the shining motif or the spiral as well as the Ak'bal sign denoting darkness (figure 13.1d–e). It is possible that the "bright" and "dark" mirrors refer to two distinct types of stone, but the distinct surfaces may allude to the quality of reflective light, with one being the bright light of day and the other, dark regions or the time of night. David Stuart (2010: 291) has recently identified a personified form of bright reflection in

Classic Maya art, a being he aptly denotes as the "shiner," describing it as an "animate mirror." Bearing traits of both the Classic Maya sun god and God C—the embodiment of sacredness—this being commonly appears on hard or polished surfaces, such as gleaming jade (figure 13.1f–h). Structure 4 from Pomona, Tabasco, had flanking balustrades of this being on its uppermost tier (figure 13.1h). Each panel had nine recessed circular elements surrounding the face, clearly for receiving disks of some foreign material, perhaps obsidian but more likely polished jade. Dating to roughly AD 300, a stucco stairway block from the North Acropolis at Tikal portrays a "dark shiner" displaying the encircling "eye cruller" of the Jaguar God of the Underworld as well as Ak'bal markings on his cheeks (figure 13.1i). It is noteworthy that for profile portrayals of mirrors in Late Classic Maya vessel scenes, the surface is always black, probably because they are usually depicted within palace rooms, which were surely dark places. Clearly, the ancient Maya were fascinated with contrasting different types of luminosity, a theme that no doubt related to their use of mirrors as well.

Whereas polished pyrite is entirely opaque, obsidian is translucent, and clearly was considered as a very different source of spiritual insight. Kovacevich also contrasts the two types of mirror stone in her chapter herein, discussing pyrite manufacture at Cancuén and noting that in the sixteenth-century *Florentine Codex* compiled by Fray Bernardino de Sahagún (1950–1982, book 11: 228), a mirror of "white stone" is for clear vision whereas that of black stone is the source of chaos and confusion:

> of these mirror stones, one is white, one black. The white one—this is a good one to look into: the mirror, the clear transparent one. They named it the mirror of the noblemen, the mirror of the ruler. The black one—this is not good. It is not good to look into; it does not make one appear good. It is one (so they say) which contends with one's face.

Although Sahagún does not specify the types of stone used for these two types of mirror, it is entirely possible that the illuminating mirror was pyrite and the other, polished obsidian. In the *Florentine* account of Quetzalcoatl leaving Tollan after being bested by Tezcatlipoca—the embodiment of the obsidian mirror—he gazes into a mirror that causes him much distress: "he called forth for his mirror. Thereupon he looked at himself; he saw himself in his mirror; he said: 'Already I am an old man'" (Sahagún 1950–1982, bk. 3: 33).

As is noted in the chapters by José Lunazzi and Olivia Kindl, one of the most striking traits of mirrors is the unique ability of an individual to see his or her own image, an obvious tool of "self-reflection." In addition, like one's

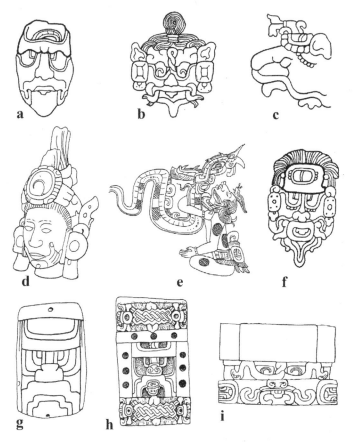

FIGURE 13.1. *Classic Maya motifs of brilliance and darkness: (a) Palenque stucco sun god head with crossed eyes and brilliance sign on brow (from Houston and Taube 2000: figure 19c); (b) face of the Jaguar God of the Underworld with spiral eyes, Palenque Tablet of the Foliated Cross (from Houston and Taube 2000: figure 19d); (c) jaguar head with mirror on brow with shining motif (from Taube 1992b: figure 13c); (d) Patron God of the numeral zero with spiral mirror on brow, Copan (from Houston and Taube 2000: figure 19g); (e) Early Classic Hero Twin with Ak'bal mirror on back, detail of Early Classic conch trumpet (from Taube 1992b: figure 13d); (f) "Shiner" deity head from the Temple of the Foliated Cross, Palenque (from Houston and Taube 2000: figure 19b); (g) "Shiner" on large jadeite bead, Palenque (from Taube 2012a: 35); (h) "Shiner" on stairway balustrade from Structure 4, Pomona (after García Moll 2005: plate 6c); (i) Early Classic "Dark Shiner" from North Acropolis, Tikal (after Coe 1990: figure 97a).*

shadow, this image is immaterial and could readily be conceived as an aspect of one's soul. Lunazzi also notes that reflections in pools of water could be used, especially if the interior of a vessel was dark. Along with being well-documented by Ruíz de Alarcón for early colonial Guerrero, the use of water vessels for divinatory scrying could be of far greater antiquity in this region. I suspect that the finely carved bowls of dark serpentine from Xochipala, Guerrero, may have served this purpose. In fact, one such bowl has half images of frogs that only can be read when copied with their opposing "mirror image" (see Gay 1972: figure 35). Similarly, many of the finely carved blackware Las Calzadas bowls from Early Formative San Lorenzo could have been used for water divination.

Ancient Mesoamerican art portrays images of men staring into mirrors with clear signs of bemusement, including an Early Formative Olmec style figurine attributed to Las Bocas, Puebla, who appears to be laughing (see Berjonneau et al. 1985: no. 42). A finely painted Late Classic Maya vase portrays a palace scene of a seated lord wearing a rabbit headdress grinning into a mirror, and it will be subsequently noted that this scene also involves inebriation and alcohol (see Blainey, chapter 9, this volume). Another Late Classic vessel scene depicts a dog enthusiastically peering into an overturned urn of alcohol while a monkey scribe gleefully gazes into a mirror while engaged in an exuberant prance-like dance, in stark contrast to the measured courtly dance exhibited by kings on monuments (see figure 13.2b). In a related scene, a monkey scribe stares raptly into a mirror as he crouches on the floor with his exposed loincloth between his buttocks, a bodily perspective very rarely encountered in Maya art (figure 13.2c).

One Late Classic Maya vase portrays an especially detailed celebratory drinking scene with pots of alcohol and enemas with a series of Chahks playing music and four aged deities, often referred to as God N, seated with their wives or consorts (see Coe 1978: Vase 11). One of the old gods grins and gesticulates before a mirror with two elements projecting from its rim (figure 13.2a). Colored green, they are a pair of earspools, although for matters of perspective placed at the lower and upper edge rather than horizontally at the center so that both can be seen. In the case of the two cited monkey-scribe mirrors, these projecting elements can also be discerned, and they are clearly earspool assemblages. The earspools denote the mirror surface as the face of the viewer while wearing the jade earpieces, thereby fluidly blending the human observer with the mirror. Portrayals of mirrors in Teotihuacan-style art from both Teotihuacan and the Escuintla region of Guatemala often appear with flanking earspools on the rim (figure 13.3a; see Taube 1992a: figures 3a, 7, 8,

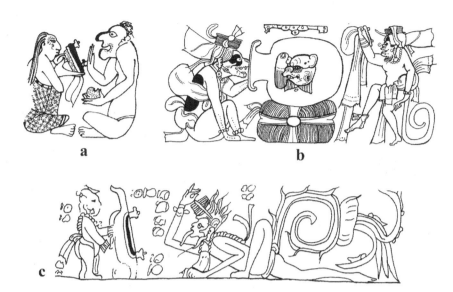

FIGURE 13.2. *Scenes of mirror use from Late Classic Maya vases (all drawings by author): (a) aged deity staring into mirror held by female companion; note earspools on rim (after Coe 1978: No. 11); (b) dog gazing into overturned vessel and monkey dancing with mirror (after Kerr 1989: 15[K505]); (c) monkey gazing into mirror held by dwarf; note earspools on rim (after www.mayavase.com K9180).*

12a, 13a, 16b). In fact, an actual Early Classic pyrite mosaic mirror with a pair of jade earspools mounted on its rim was excavated at Kaminaljuyu (see Kidder et al. 1946: figure 6c). In addition, human faces are at times portrayed within mirrors, with perhaps the most striking examples being the *tezcacuitlapilli* turquoise-back mirrors worn by the great atlantean columns from Building B at Tula (figure 13.5d). In this case, the center where the pyrite mosaic would be is probably the shining reflective face of the sun god.

Rather than passive and inert objects, ancient Mesoamerican mirrors were surely considered to be sources of information with stories to tell, much like reading a sacred book: "[t]o see oneself in a mirror was equivalent to reading a book of destinies, or rather this kind of book was identified with mirrors" (Olivier 2003: 256). The K'iche' *Popol Vuh* refers to the original, preconquest manuscript as an *il'bal*—an instrument for "seeing," the same term that modern K'iche' use for such items as divinatory crystals or telescopes (see McGraw, chapter 10, this volume; Tedlock 1996: 218). Tedlock (ibid.) notes that the *Popol Vuh* was a magical book revealing mythological doings by the gods and

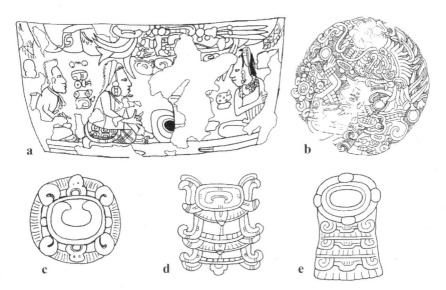

FIGURE 13.3. *The Principal Bird Deity and Classic Maya avian imagery pertaining to mirrors (all drawings by author): (a) Terminal Classic vessel scene with mirror and hovering Principal Bird Deity from the Sacred Cenote, Chichen Itza (after Anonymous 2008: 51); (b) Principal Bird deity atop tree, detail of Early Classic slate mirror back (from Taube et al. 2010: figure 34c); (c) mirror with lunar sign worn on back of Principal Bird Deity, detail of Early Classic ceramic vessel (after Coe 1989: figure 14); (d) back mirror with probably lunar crescent and pendant bird tail from Early Classic vessel portraying Principal Bird Deity (after Hellmuth 1987: figure 494); (e) mirror with pendant bird tail worn by Early Classic deity, detail of two-piece effigy vessel from Burial 10, Tikal (from Taube 1992b: figure 36b).*

such future events as war and famine, topics probably addressed to mirrors as well. Ruíz de Alarcón recorded an early seventeenth-century Nahuatl chant describing the act of water scrying as looking into a book or a mirror (Coe and Whittaker 1982: 214). In addition, García-Zambrano (1994: 221) notes that many Mesoamerican colonial texts refer to maps as *espejos*, or "mirrors," including the *Códice de Cholula*: "Lords and noblemen, here are your papers, the mirror of your antiquity and the history of your ancestors."

In a discussion of contemporary mirror divination among the Huichol of Santa Catarina Cuexcomatitlan, Blosser (2000: 4) notes that according to one Huichol informant "mirrors are like the apprentice's notebook: what the student learns is somehow inscribed in the mirror." In addition, Huichol

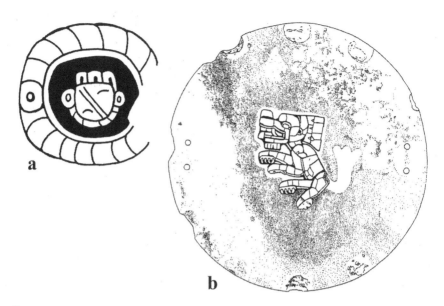

FIGURE 13.4. *Images appearing on Early Classic Teotihuacan-style mirrors (drawings by author); (a) Teotihuacan portrayal of schematic face atop mirror surface, detail of mural from Techinantitla, Teotihuacan (from Taube 1992a: figure 7); (b) Early Classic Teotihuacan-style mirror with jade and shell mosaic, Art Institute of Chicago.*

divinatory mirrors also record the sacred shrines visited by apprentice priestly shamans, or *mara'akame*:

> Apprentices carry mirrors with them whenever they visit a sacred place. What they learn as a result of the visit is somehow inscribed in the mirror. Furthermore, the mirror serves the apprentice as a channel of communication with the deity of that location.

In other words, the Huichol divinatory mirror is much like a camera that records the sacred nature of particular shrines, or perhaps a more apt analogy is that it functions similar to the sight and minds of human beings, with the images recorded in the "memory" of the object.

Clearly enough, sacred mirrors in ancient Mesoamerica were not "read" as script but as images, which is also known for Greco-Roman traditions of mirror divination, as in the following account by Pausanias describing a second-century Greek temple ritual: "If anyone looks into this mirror, he will see himself very dimly or not at all, but the actual images of the gods . . . can be seen quite clearly" (Addey 2007: 36). Among the contemporary Huichol there is the

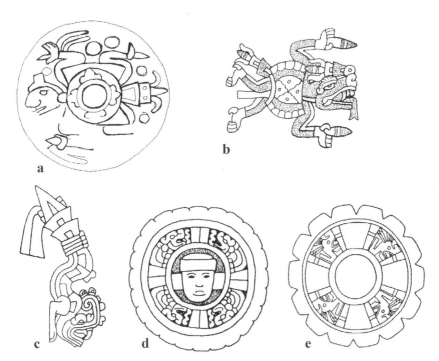

FIGURE 13.5. *Turquoise and pyrite mirror disks and the Mixtec Yahui and Central Mexican Xiuhcoatl, or "turquoise fire serpent" (all drawings by author); (a) Mixtec-style turquoise mosaic disk of Yahui figure found near Acapulco, coastal Guerrero (after Bernal 1951); (b) Flying Yahui with flint blades in hands, compare with figure 5a, Codex Nuttall, page 12 (from Taube 2012b: figure 4d); (c) detail of Aztec monument portraying diving Xiuhcoatl, British Museum, London (from Taube 2012b: figure 4e); (d) Toltec portrayal of turquoise and pyrite mosaic back mirror, or tezcacuitlapilli, with face on center mirror surface and four Xiuhcoatl serpents on turquoise rim, detail of Atlantean warrior column from Tula, Hidalgo (from Taube 1992a: figure 12c); (e) schematic drawing of Toltec-style pyrite and turquoise mirror excavated at Chichen Itza, Yucatan (from Taube 1994: figure 31a).*

concept of *nearika*, or an instrument of seeing, which can refer to sacred art, including traditional yarn paintings, as well as circular glass mirrors, all of which provide access to the realm of the ancestors (see Kindl, chapter 12, this volume).

The concept of seeing figures and images in mirrors is widespread in ancient Mesoamerica. One of the ill omens witnessed by the Aztec just before the coming of the Spanish was a bird with a mirror portraying images in vivid detail:

Its crest was like a round mirror pierced in the center like one a god might use to look into the future. There Moctezuma peered, to see the heavens—the stars—the fire drill constellation. He was first startled, and then terrified, as he saw, a little beyond, what looked like fighting men massed, like conquerors in war array, riding the backs of deer. (Sahagún 1950–1982, bk. 12)

In a recent study, Tomás Villa Córdova (2010: 114–115) mentions a number of early colonial accounts describing mirror visions, including not only the text cited immediately above, but also a similar account of the Spanish conquest, here concerning the last Aztec emperor Cuauhtemoc, who saw Aztec citizens (*macehualtin, hombres del pueblo)* in the mirror surface, a sign that they were to be defeated. As discussed by Marc Blainey in chapter 9, a number of Classic Maya vessels graphically depict visionary scenes concerning mirrors with clearly supernatural creatures, including a palace with an odd, heron-like bird and a host of plump, gamboling bunnies (see Kerr 1990: 205 [K2026]). Still another vessel scene has a human figure with a creature combining the traits of both a jaguar and a rabbit (see chapter 9: figure 9.1). What is noteworthy is that these beings are *outside* the mirror, a blending of the human and supernatural. These scenes suggest that mirrors served as cave-like portals both to and *from* the numinous realm of gods and ancestors, a theme also discussed by Kindl in chapter 12. In addition to the examples cited by Blainey, a Terminal Classic stucco-painted vessel from the Sacred Cenote at Chichen Itza portrays a mirror with another supernatural creature, in this case the Principal Bird Deity, who hovers directly above the mirror in a curtained palace setting (figure 13.3a). An ancestral form of Vucub Caquix of the *Popol Vuh*, the Principal Bird Deity is now known to be an avian aspect of the aged creator god Itzamnaaj, and the preeminent bird messenger, or *mut* for auguries (Houston et al. 2006: 234–241). An Early Classic slate mirror back excavated at Zaculeu, Guatemala, portrays the Principal Bird Deity with outstretched wings, and it is likely that another more fragmentary mirror back from the same tomb portrayed a similar theme with at least one example of this being (see Woodbury and Trik 1953: figs, 131, 282a). Still another Early Classic slate mirror back depicts the Principal Bird Deity perched in a gourd tree (see figure 13.3b). In addition, this being can also appear wearing a mirror at the base of its tail, much like back mirrors worn by human figures in Classic and Postclassic Mesoamerica (figure 13.3c–d). In Classic Maya iconography, back mirrors often have the bird tail of the Principal Bird Deity, including a well-known Early Classic effigy vessel of a seated old god from Tikal (figure 13.4c–d).

Although some carved or painted mirror backs could refer to personal ownership, many may denote the types of visions seen in the mirror surface. Mention was made above of a probable image of a solar face on the Atlantean column back mirrors at Tula (figure 13.5d), and similarly, a Teotihuacan mural portrays a stylized face in the center of the reflective surface (figure 13.4a). A remarkable Early Classic mirror at the Art Institute of Chicago portrays a large pyrite face with no mosaic tesserae, probably an example of the aforementioned "composite" mosaic of pyrite fragments polished after being glued onto a supportive matrix. The center of this pyrite face features a pouncing feline rendered in jade and shell mosaic, and perhaps this concerns the typical placing of an image on the slate back of the mirror to the shining surface itself (figure 13.4). It is important to note that Mesoamerican pyrite mirrors have a strongly temporal component, and they are especially prevalent in the Early Classic period, that is, during the apogee of Teotihuacan. In chapter 5, Julie Gazzola and Sergio Gómez Chávez note that the remains of more than 200 slate and pyrite mirrors have been recently discovered during the current excavations in the tunnel underlying the Temple of Quetzalcoatl, a truly remarkable amount, as is the still more recent discovery of hundreds of pyrite spheres in the same tunnel. In contrast, the fashioning of pyrite mosaic mirrors is little known for the Late Postclassic period, although there are many examples of shield-like mosaic disks rendered in polished turquoise and shell mosaic, at times with complex supernatural scenes (see Saville 1922). Often rendered in differing shades of turquoise, the mosaic scenes on the turquoise disk pieces are frequently difficult to "read" without effort, much as if one were peering into a mirror and seeing fleeting images of remote and ancient beings. A Late Postclassic wooden mosaic disk from the coastal Acapulco region of Guerrero, features a flying figure with a sun sign on its back (figure 13.5a). This being can be identified as the Mixtec Yahui, the necromancer *par excellence* who could even fly through stone to create contact with the supernatural realm (figure 13.5b; see Taube 2001b). In other words, much like the scenes of the Principal Bird Deity on the backs of Early Classic Maya mirrors, this being serves as a supernatural contact to the spirit world. It is well known that the Yahui is a Mixtec version of the Aztec Xiuhcoatl, or turquoise fire serpent, and forms of this being appear on the turquoise rims of Toltec-style mirrors, containing a central pyrite mosaic disk (figure 13.5d–e). Dating to the Early Postclassic, such mirrors with their broad turquoise rims and relatively small pyrite centers may constitute a transitional link between Classic-period pyrite mirrors and the turquoise mosaic disks of the Late Postclassic.

In his treatise concerning Classic Maya mirror scenes, Blainey (chapter 9, this volume) notes that the Classic Maya may have been using hallucinogens to induce shamanic visions with mirrors. Although this is a possibility, dreaming is another form of altered state that can provide surreally beautiful or frightening concepts, episodes, and images, and is an important form of divination and prognostication among the Maya today (see Bruce 1975; Laughlin 1976). As aptly put by Robert Laughlin (1976: 3):

> Dogs dream, and cats dream. Horses dream, and even pigs, say the Zinacantecs. No one knows why; but there is no question in the mind of a Zinacantec why men dream. They dream to live a full life. They dream to save their lives.

In addition, in Mayan languages, the term for sleeping or dreaming, *way*, also relates to one's spiritual co-essence, which in Classic Maya vessel scenes are frequently quite spooky beings appearing as supernatural animals as well as gods of death and disease (Houston and Stuart 1989). A 1704 Spanish account from the Suchitepequez region of southern Guatemala describes native priests using stone mirrors to determine the spiritual co-essence of children, although here referred to by the related and overlapping term *nahual:* "the quality of the Nahual that they showed to their children and which was the animal that first appeared in that stone, and the healers and physicians could see there their patients and the issue of their illness" (translation in Olivier 2003: 256). In her discussion of divinatory mirrors among the contemporary Huichol in chapter 12, Kindl notes that rather than presenting visions or images of the ancestors, their voices are heard through the mirror and also appear subsequently in dreams. Similarly, a sixteenth-century account for Tezcoco mentions that while the mirror of Tezcatlipoca could speak to them during their migration, it only communicated through the dreams of priests once the city was founded.

Aside from dreams, alcoholic inebriation constitutes an extremely important form of altered state in traditional Mesoamerica, and rather than being a profane form of enjoyment, it is a major component of sacred ritual behavior for contact with the otherworldly realm of gods and ancestors. Among contemporary Zinacantecos of highland Chiapas, distilled cane liquor, or *pox*, is referred to as "dew-drops of the gods" (Vogt 1976: 35). The essence of this beverage concerns the soul and the spirit world, as well as social interactions in the world of human beings:

> POX has a "hot" and "strong" innate soul which serves to open channels of communication and reduce noise and distortion which might interfere with

transactions between men or between men and the gods. The flow of POX in Zinacanteco society, if it could be measured and charted, would provide a blueprint and mirror of social relationships. (Vogt 1976: 36)

Not only does this discussion of the sacred use of *pox* bring up previous themes of mirrors as something to be read, such as a map-like blueprint, but also as a "mirror of social relationships," a theme very much related to mirrors and concepts of rulership and governance in ancient Mesoamerica (see Kovacevich, this volume, chapter 4). However, how can any public political event, social inebriation, or the fleeting glimmer of a mirror surface be "read" and evaluated without the subsequently "reflective" and now passive concept of memory? The close relation of alcohol and inebriation with contact to the spirit world accords well with many portrayals of *way* beings on Late Classic Maya vases, beings who enthusiastically consume drink from ceramic *ollas* as well as handling enema syringes for imbibing the substance through other means. Many of these vessels are marked glyphically as *chih*, a term that probably refers to the fermented sap of the agave, or *pulque* in contemporary highland Mexico (for a recent discussion of pulque, alcohol, and inebriation among the Classic Maya, see Houston et al. 2006: 116–122).

In terms of the relation of *way* spirits and contact with the realm of gods and ancestors in Maya thought, it is worth mentioning again the mirror scene of Chahks, quadripartite aspects of God N and their female companions, with one God N happily staring into a mirror (figure 13.2a). In another previously mentioned vessel scene, a corpulent lord with a rabbit headdress sits on his throne while a dwarf drinks from a massive bowl near an adjacent jug of alcohol (see Kerr 1989: 86 [K1453]). However, an especially intriguing scene of mirror use and alcohol is the aforementioned depiction of a monkey scribe with writing quills in his headdress while happily dancing before a mirror carried in his raised hand (figure 13.2b). On the other side of the vessel, a dog gazes with great enthusiasm into an overturned urn of alcohol. In terms of the flaring rims of the mirror bowl and that of the urn, the connection of both being instruments of "seeing" appears to be intentional. As has been mentioned in many of the chapters in this volume, liquid surfaces served as a form of divinatory scrying. This is also true for ancient Greco-Roman traditions. In a fresco from the Villa of the Mysteries in Pompeii, a satyr gazes intently into a vessel held by Silenus, the drunken diviner companion of Dionysus, the god of wine (see Addey 2007: figure 2.1). As noted by Addey (ibid.: 37), "[m]irrors were certainly linked with the mystery ceremonies in honor of Dionysus." Rather than hallucinogens, alcohol was the means of acquiring special visions in ancient

Greece; wine was "a way of being moved transitorily to a level above daily life: to see and also reveal reality beyond appearances" (Isler-Kerényi 2007: 233). Classic Maya vessel scenes indicate that along with dreaming, drinking and intoxication served as means of contacting the spirit world through the use of mirrors.

Mention has been made of a Tezcocan account of the speaking mirror of Tezcalipoca, and despite being instruments of seeing, divinatory mirrors and related objects are often described as having aural qualities, perhaps relating to the immaterial and ethereal nature of souls and spirit beings. Describing the *nierika* mirror as a means of communicating with gods, one Huichol informant stated that "[t]he *mara'akame's* telephone is *nierika*" (Blosser 2000: 5). Similarly, in chapter 10 John McGraw mentions that divinatory crystals have been compared to a radio among contemporary highland Maya. Among the Classic Maya, there truly was a "speaking mirror." The great king of Palenque, K'inich Janahb Pakal, was buried with a curious item over his face, this being a rectangular pyrite-and-shell mosaic plaque that framed his mouth (see Ruz Lhuiller 1973: 255–256, figures 220–221). As Ruz Lhuiller notes, the same object is worn by the stucco figures appearing on the tomb walls of the Temple of the Inscriptions (figures 176, 178). In addition, it can be seen with similar but earlier figures painted on the walls of the Temple XX subtomb (see Robertson 2001). On the sarcophagus lid, Pakal is clearly portrayed as the Maya maize god; the mirror mouthpiece is closely related to this deity and appears with him on a Late Classic vase, in this case with a floral breath element emerging from the center (figure 13.6a). Moreover, it is worn by the ruler on Dos Pilas Stela 17, who impersonates the maize god (figure 13.6b). In the scene, he wears a massive, thick belt that clearly alludes to a royal throne, and the rectangular mirror plaque is at times portrayed on thrones with versions of floral breath rendered in jade emanating from the center, much as if it is speaking (figure 13.6c). Temple 18 at Copan featured a cornice in the form of a mat motif with the mirror device and probable renderings of jade masks, suggesting that the structure was a symbolic "throne house" (figure 13.6d; see Taube 2012a: 39). The relation of the rectangular plaque to thrones and kingship is apt, considering the strong relation of mirrors to rulers in both Central Mexico and the Maya area (see Kovacevich, chapter 4, this volume).

In chapter 9, Marc Blainey discusses the concept of a "reflective surface complex" for ancient Mesoamerica, pertaining to luminosity in relation to divinity and visions (see also Healy and Blainey 2011). In my view, this very much pertains to the Flower World complex first defined by linguist Jane Hill (1992) for Mesoamerica and the Greater Southwest, who noted that it is

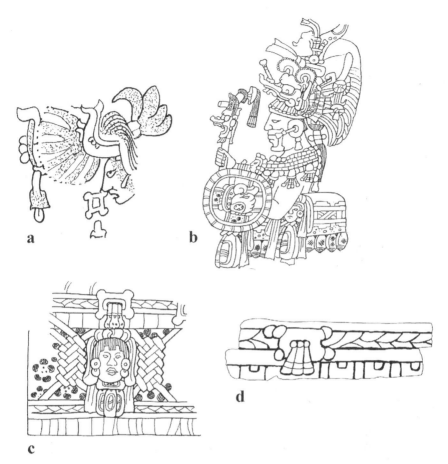

FIGURE 13.6. *The pierced four-cornered mirror motif in Late Classic Maya art (all drawings by author); (a) Maya maize god with quadrangular element before mouth (after mayavase.com.K5615); (b) Dos Pilas ruler wearing mirror device on face, note throne "belt" as well (from Taube 2005: fig. 7d); (c) throne with four-sided mirror element, Lintel 3, Temple 1, Tikal (from Taube 2012: 38); (d) throne cornice with quadrangular mirror element, Temple 18, Copan (from Taube 2012: 39).*

a shining solar realm pertaining to gleaming stones as well as precious birds, butterflies, and flowers. Louise Burkhart provided a vivid description of Aztec concepts of this supernatural realm:

> The garden is a shimmering place filled with divine fire; the light of the sun reflects from the petals of the flowers and the iridescent feathers of birds;

human beings—the souls of the dead or the ritually transformed living—are themselves flowers, birds, and shimmering gems. (Burkhart 1992: 88–109)

I would suggest that in ancient Mesoamerica, circular mirrors also alluded to flowers, as I have previously noted for ancient Teotihuacan, where they are frequently portrayed with petalled rims (Taube 1992a). A stucco-painted Teotihuacan-style mirror back from Kaminaljuyu displays a series of flowers on its rim (see Kidder et al. 1946: figure 205b), and similarly, a carved Teotihuacan style mirror back from Michoacan features four explicit flowers on its rim and another in the collection of the Denver Art Museum portrays butterfly elements, with butterflies being obvious denizens of the Flower World (figure 13.7d–e). In addition, a mural from Techinantitla depicts a cleft blossom emerging from the mirror surface, a motif virtually identical to flowers from the murals at Tepantitla (figure 13.7a–b). It is also important to note that in Mesoamerica and the Greater Southwest, yellow is the prevalent color of flowers, surely alluding to the sun as well as such flowers as squash blossoms, sunflowers, and marigolds. Clearly enough, yellow is also the color of golden pyrite mirrors, which were surely compared to flowers as well as the sun.

Many authors in this volume noted that in Mesoamerica, mirrors serve as cave-like passageways for supernatural beings (see also Taube 1992a; Villa Córdova 2010). Much like Lewis Carroll's *Through the Looking-Glass*, mirrors in ancient Mesoamerica marked a sharp distinction between the world of humans and the other world, with no possibility of our fleshy human hands protruding through the shining surface. As reflecting passageways, ancient Mesoamerican mirrors were probably thought of not simply as solid shiny surfaces, but as "holes" penetrating to the other side. Thus for the Huichol, not only can a circular mirror signify the sun, but also a simple circular frame around an open orifice serves the same meaning (see Berrin 1978: plates 13, 16). As Juan Negrín (1975: 19) notes while discussing Huichol "front shield" yarn and stick *nierika*, "The hole in the center seems more like an opening into the supernatural world, a channel through which the Ancestors may enter into man's reality and the shaman may gaze into their transcendental realm." Clearly an open hole allows light to pass directly through the object rather than reflecting it, with the most important source of light being the sun. In this context, it is noteworthy that Joseph Mountjoy has excavated Middle Formative pyrite plaques from El Pantano, Jalisco, with holes drilled through the center, suggesting that the Huichol concept of mirrors as holes or portals is of great antiquity in the Jalisco region (see chapter 7 as well as figure 13.6). Nicholas Saunders (2003: 17) notes that there is a widespread

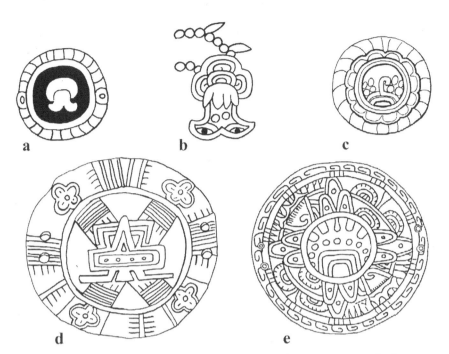

FIGURE 13.7. *Teotihuacan mirrors and flower symbolism (all drawings by author):*
(a) Teotihuacan mural portrayal of mirror with central blossom and flanking earspools
(from Taube 1992a: figure 13a); (b) Teotihuacan depiction of flower, detail of mural from
Tepantitla, Teotihuacan (from Taube 1992a: figure 13b); (c) Teotihuacan ceramic "adorno"
of "Reptile's Eye" motif within petalled mirror rim (from Taube 1992a: figure 13e); (d)
small slate mirror back from Quitzeo Basin of Michoacan with four flowers on rim (after
Filini 2004: figure 5.17); (e) Teotihuacan-style slate mirror back with four mountains and
butterfly elements surrounding central "Reptile's Eye" sign, Denver Art Museum (from
Taube 2009: figure 1a).

association between brightness and the spirit world among New World peoples: "Despite expected and complex differences among Amerindian outlooks,
varying attitudes toward brilliant objects appear to have emerged from and
cohered around a worldview that saw light, dazzling colors, and shiny matter as indicating the presence of supernatural beings and essence." In chapter
12 of the present volume, Olivia Kindl notes the Huichol concept of crystals
being "glimmering condensations of the ancestors." In addition, she mentions
the Huichol belief of living human souls being shining celestial drops of water
that evaporate at death.

As is indicated throughout this volume, mirrors, polished jade, and even dewdrops at dawn relate to the Mesoamerican concept of shining elements being windows or passageways for souls and gods. The preeminent source of this "reflective surface complex" is clearly the sun. Although in contemporary Western thought it is assumed that the sun is a globe, there is no indication that this was this case in ancient Mesoamerican thought. Instead, ancient Mesoamerican imagery indicates that it was a small, brilliant hole of light that traveled the sky on a daily basis, thereby offering access to a celestial realm of spirit beings. The obvious corollary is that at night, it was a "dark hole" offering contact with surely less attractive denizens of the underworld. In Maya art, the sun god and ancestors can appear as projecting out of solar mirrors, and the upper registers of the Temple of the Warriors columns at Chichen Itza feature the celestial sun god emerging out of an open hole surrounded by a solar rim (figure 13.8a–c). The side of an Aztec ceramic drum features a solar disk with a central orifice from which the sound would exit, music being another basic means to contact the spirit world in Mesoamerica (see www. mayavase.com.K6021). The concept of a solar mirror portal may be of great antiquity, and mention has been made of the perforated pyrite plaques from El Patano, Jalisco. The roughly contemporaneous Altar 4 from the Olmec site of La Venta, Tabasco, features a muscular male figure in a niche surrounded by a round petalled rim, much like a mirror worn on the brow of a jadeite statue attributed to Arroyo Pesquero (figure 13.8d; see Taube 2004b: figure 48). Although this scene has been widely interpreted as a ruler emerging out of a cave, why cannot it be a celestial mirror portal? The figure wears a feathered bird headdress and has similar feather-like elements projecting from his shoulders suggestive of flames. Although it is by no means clear that this is an Olmec sun deity, mural fragments from Pinturas Sub–1a at San Bartolo indicate that by at least the first century BC, the Maya conceived of an avian sun god (see Taube et al. 2010: figure 21c).

The La Venta Altar 4 disk is surrounded by four elements that I have previously identified as celtiform maize ears, a common way of delineating the four directions and cosmos among the Olmec (Taube 2000: 301–303). Similarly, an aforementioned Teotihuacan-style mirror back featuring butterfly imagery also has four mountains to the cardinal points, suggesting that it encompasses the entire world (figure 13.7d). It is readily apparent that many Classic-period mirrors have four elements on the rim, thereby framing them with cosmic directional symbolism (figures 13.1d, 13.3c, 13.3e, 13.5d–e, 13.6d). Olivia Kindl (chapter 12, this volume) notes that among the Huichol, mirrors can denote the earth's surface and the entire world, a concept that I have also discussed

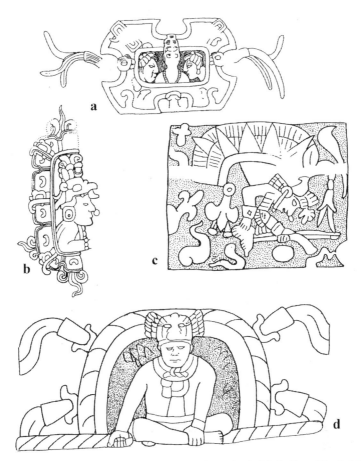

FIGURE 13.8. *The sun as a "portal" (all drawings by author): (a) the Late Classic Maya sun god with royal ancestors emerging from a solar disk flanked by precious birds, Palenque (from Taube 2004a: figure 8c); (b) royal ancestor emerging from solar mirror, detail of Caracol Stela 5 (from Taube 1992a: figure 21a); (c) sun god emerging from solar portal with flames and flowers, detail of Column 55, Northwest Collonade, Chichen Itza (after Morris et al. 1931: plate 118); (d) Olmec avian figure in feather-rimmed portal, La Venta Altar 4 (drawing after photograph by author).*

for ancient Mesoamerica (Taube 1992a). This relates to the preeminent Aztec seeing instrument of Tezcatlipoca known as the *tlachianoli,* or "viewer," a spoked disk on a handheld scepter (13.9). This item is so closely identified with Tezcatlipoca that it serves as the Aztec glyphic sign for the *veintena* month Toxcatl, dedicated to this god (13.9b). Although mirror-like in quality,

FIGURE 13.9. *The tlachianoli, or "viewer," a spoked disk on a handheld scepter; (a) from the Codex Magliabechiano; (b) the Aztec glyphic sign for the veintena month Toxcatl.*

the center of the disk was an open hole, as can be graphically seen in the *Codex Magliabechiano*, where the supporting stick can still be seen through the center (13.9a). The *Primeros Memoriales* account of the regalia of Tezcatlipoca describes him grasping his *tlachianoli*: "In his other hand is the viewer (*tlachianoli*); it has a hole through which he watches people" (Sahagún 1997: 95). In a similar description of this god's accoutrements, Diego Durán (1971: 99) also mentions this seeing device, but in this case as a golden mirror, "a round plate of gold, shining and brilliant, polished like a mirror. This [mirror] indicated that Tezcatlipoca could see all that took place in the world with that reflection." Clearly, as in the case of the contemporary Huichol, there was considerable overlap between mirrors and open passageways in Aztec thought.

As his instrument of seeing, the *tlachianoli* was clearly related to the Aztec concept of Tezcatlipoca being omnipresent and omniscient, as can be seen in the following prayer in which he is addressed:

> O master, O our lord, O lord of the near, of the nigh: thou seest, thou knowest the things within the trees, the rocks. And behold now, it is true that thou knowest of things within us; thou hearest us from within. (Sahagún 1950–1982, bk. 6: 25)

As noted above, the concept of "hearing" as well as seeing can relate to Mesoamerican concepts of divinatory scrying. The all-knowing abilities of Tezcatlipoca recall the creation of the present race of humans in the K'ichean *Popol Vuh*. In this episode, the gods blurred the vision of the recently created humans: "[t]hey were blinded as the face of a mirror is breathed on" (Tedlock 1996: 148). To turn it the other way, what was their mirror-like vision like before it became misted?:

> Perfectly they saw, perfectly they knew everything under the sky, whenever they looked. The moment they turned around and around in the sky, on the earth, everything was seen without obstruction. They didn't have to walk around before they could see what was under the sky; they just stayed where they were.
>
> As they looked, their knowledge became intense. Their sight passed through trees, through rocks, through lakes, through seas, through mountains, through plains. (Tedlock 1996: 147)

Not only do the Aztec and K'iche describe the same ability to see through trees and rocks, but both relate to mirrors, with the *Popol Vuh* alluding to the vision offered by a shining, unblurred mirror, and Tezcaltipoca's name signifying "Smoking Mirror." Although after centuries of time and "misting"—the oxidization of most ancient mirrors—such accounts offer remarkably vivid insights into the truly magic nature of these fascinating and once resplendent objects.

BIBLIOGRAPHY

Acosta, Jorgé. 1964. *El palacio de Quetzalpapalotl*. Mexico City: Instituto Nacional de Antropología e Historia.

Addey, Crystal. 2007. "Mirrors and Divination: Catoptromancy, Oracles and Earth Goddesses in Antiquity." In *The Book of the Mirror*, ed. Miranda Anderson, 32–46. Newcastle: Cambridge Scholars Publishing.

Anonymous. 2008. "Guía visual Chichén Itzá Yucatán." *Arqueología Mexicana,* edición especial no. 27.

Berjonneau, Gerald, Emile Deletaille, and Jean-Louis Sonnery. 1985. *Rediscovered Masterpieces of Mesoamerica: Mexico-Guatemala-Honduras.* Bologne: Editions Arts.

Bernal, Ignacio. 1951. "Nuevos descubrimientos en Acapulco, Mexico." In *The Civilizations of Ancient America: Selected Papers of the XXIXth International Congress of Americanists,* edited by Sol Tax, 52–56. Chicago: University of Chicago Press.

Berrin, Kathleen, ed. 1978. *Art of the Huichol Indians.* New York: Harry N. Abrams.

Blainey, Marc. 2007. "Surfaces and Beyond: The Political, Ideological, and Economic Significance of Ancient Maya Iron-Ore Mirrors." Master's Thesis, Department of Anthropology, Trent University.

Blosser, Bret. 2000. "Alternative Constructions of Religious Expertise in a Huichol Community, Jalisco, Mexico." www.famsi.org/reports/95114.

Boone, Elizabeth, ed. 1996. *Art at Dumbarton Oaks.* 2 vols. Washington, DC: Dumbarton Oaks.

Bruce, Robert D. 1975. *Lacandon Dream Symbolism.* Mexico City: Ediciones Euroamericanas Klaus Thiele.

Burkhart, Louise. 1992. "Flowery Heaven: The Aesthetic of Paradise in Nahuatl Devotional Literature." *Res: Anthropology and Aesthetics* 21: 88–109.

Carlson, John B. 1981. "Olmec Concave Iron-Ore Mirrors: The Aesthetics of a Lithic Technology and the Lord of the Mirror." In *The Olmec and Their Neighbors: Essays in Memory of Matthew W. Stirling,* ed. Elizabeth P. Benson, 117–147. Washington, DC: Dumbarton Oaks.

Coe, Michael D. 1978. *Lords of the Underworld: Masterpieces of Classic Maya Ceramics.* Princeton: Princeton University Press.

Coe, Michael D. 1989. "The Hero Twins: Myth and Image." In *The Maya Vase Book,* vol. 1. ed. Justin Kerr, 161–184. New York: Kerr Associates.

Coe, Michael D., and Gordon Whittaker. 1982. *Aztec Sorcerers in Seventeenth Century Mexico: The Treatise on Superstitions by Hernando Ruiz de Alarcón.* Institute for Mesoamerican Studies, Pub. 7. Albany: State University of New York at Albany.

Coe, William. 1990. *Excavations in the Great Plaza, North Terrace and North Acropolis of Tikal.* Tikal Report No. 14. Philadelphia: University Museum, University of Pennsylvania.

Durán, Diego. 1971. *Book of the Gods and Rites and the Ancient Calendar.* Norman: University of Oklahoma Press.

Easby, Elizabeth Kennedy, and John F. Scott. 1970. *Cortés: Sculpture of Middle America.* New York: New York Graphic Society.

Evans, Susan Toby, ed. 2010. *Ancient Mexican Art at Dumbarton Oaks*. Washington, DC: Dumbarton Oaks.

Filini, Agapi. 2004. *The Presence of Teotihuacan in the Cuitzeo Basin, Michoacán, Mexico*. BAR International Series 1279. Oxford: British Archaeological Reports.

García-Moll, Roberto. 2005. *Pomoná: Un sitio del Clásico Maya en las colinas tabasque-ñas*. Mexico City: Instituto Nacional de Antropología de Historia.

García-Zambrano, Angel J. 1994. "Early Colonial Evidence of Pre-Columbian Rituals of Foundation." In *Seventh Palenque Round Table, 1989*, ed. Merle Greene Roberston, 217–227. San Francisco: Pre-Columbian Art Research Institute.

Gay, Carlo T. E. 1972. *Xochipala: The Beginnings of Olmec Art*. Princeton: Art Museum, Princeton University.

Gladwin, Harold S., Emil W. Haury, E. B. Sayles, and Nora Gladwin. 1938. *Excavations at Snaketown: Material Culture*. Medallion Papers no. 25. Gila Pueblo: Globe.

Healy, Paul F., and Marc G. Blainey. 2011. "Ancient Maya Mosaic Mirrors: Function, Symbolism, and Meaning." *Ancient Mesoamerica* 22 (2): 229–244. http://dx.doi.org/10.1017/S0956536111000241.

Hellmuth, Nicholas. 1987. *Monster und Menschen in der Maya-Kunst*. Graz: Akademische Druck, u. Verlagsanstalt.

Hill, Jane H. 1992. "The Flower World of Old Uto-Aztecan." *Journal of Anthropological Research* 48: 117–144.

Houston, Stephen, and David Stuart. 1989. "The Way Glyph: Evidence for 'Co-essences' among the Classic Maya." *Research Reports on Ancient Maya Writing* 30. Washington, DC: Center for Maya Research.

Houston, Stephen, and Karl Taube. 2000. "An Archaeology of the Senses: Perception and Cultural Expression in Ancient Mesoamerica." *Cambridge Archaeological Journal* 10 (2): 261–294. http://dx.doi.org/10.1017/S095977430000010X.

Houston, Stephen, David Stuart, and Karl Taube. 2006. *The Memory of Bones: Body, Being, and Experience among the Classic Maya*. Austin: University of Texas Press.

Isler-Kerényi, Cornelia. 2007. *Dionysos in Archaic Greece: An Understanding through Images*. Trans. Wilfred G. E. Watson. Leiden: Brill.

Kerr, Justin. 1989. *The Maya Vase Book*. vol. 1. New York: Kerr Associates.

Kerr, Justin. 1990. *The Maya Vase Book*. vol. 2. New York: Kerr Associates.

Kidder, Alfred V., Jesse D. Jennings, and Edwin M. Shook. 1946. *Excavations at Kaminaljuyu, Guatemala*. Washington, DC: Carnegie Institution of Washington, Pub. 561.

Laughlin, Robert. 1976. *Of Wonders Wild and New*. Smithsonian Contributions to Anthropology 22. Washington, DC: Smithsonian Institution Press.

Morris, Earl H., Jean Charlot, and Ann Axtel Morris. 1931. *The Temple of the Warriors at Chichen Itzá, Yucatan.* Carnegie Institution of Washington Pub. 406. Washington, DC: Carnegie Institution of Washington.

Mountjoy, Joseph. 1998. "The Evolution of Complex Societies in West Mexico: A Comparative Perspective." In *Ancient West Mexico: Art and Archaeology of an Unknown Past,* ed. Richard Townsend, 250–265. London: Thames and Hudson.

Negrín, Juan. 1975. *The Huichol Creation of the World.* Sacramento: E. B. Crocker Art Gallery.

Nelson, Zachary, Barry Scheetz, Guillermo Mata Amado, and Antonio Prado. 2009. "Composite Mirrors of the Ancient Maya: Ostentatious Production and Precolumbian Fraud." *PARI Journal* 9 (4): 1–7.

Nicholson, Henry B., and Rainer Berger. 1968. *Two Aztec Wood Idols: Iconographic and Chronologic Analysis.* Studies in Pre-Columbian Art and Archaeology, 5. Washington, DC: Dumbarton Oaks.

Olivier, Guilhem. 2003. *Mockeries and Metamorphoses of an Aztec God Tezcatlipoca, "Lord of the Smoking Mirror.* Boulder: University Press of Colorado.

Robertson, Merle Greene. 2001. "Los murales de la tumba del Templo XX sub de Palenque." In *La Pintura Mural Prehispánica en México, II, Área Maya, tomo IV, Estudios,* ed. Leticia Staines Cicero, 381–388. Mexico City: Universidad Nacional Autónoma de México.

Ruz Lhuiller, Alberto. 1973. *El Templo de las Inscripciones, Palenque.* Mexico City: Instituto Nacional de Antropología e Historia.

Sahagún, Fray Bernardino. 1950–1982. *Florentine Codex: General History of the Things of New Spain.* Translated by A.J.O. Anderson and C.E. Dibble. Santa Fe: School of American Research.

Sahagún, Fray Bernardino. 1997. *Primeros Memoriales: Paleography of Nahuatl Text and English Translation.* Norman: University of Oklahoma Press.

Saunders, Nicholas. 2003. "'Catching the Light': Techologies of Power and Enchantment in Pre- Columbian Goldworking." In *Gold and Power in Ancient Costa Rica, Panama and Colombia,* ed. Jeffrey Quilter and John Hoopes, 15–47. Washington, DC: Dumbarton Oaks.

Saville, Marshall H. 1922. *Turquoise Mosaic Art in Ancient Mexico.* Contributions of the Museum of the American Indian, Heye Foundation 6. New York: Heye Foundation.

Schele, Linda, and Mary Ellen Miller. 1986. *The Blood of Kings: Dynasty and Ritual in Maya Art.* New York: George Braziller.

Schmidt, Peter. 1998. *Maya.* Ed. Mercedes de la Garza and Enrique Nalda. New York: Rizzoli.

Schöndube B., Otto, and L. Javier Galvan V. 1978. "Salvage Archaeology at El Grillo-Tabachines, Zapopan, Jalisco, Mexico." In *Across the Chichimec Sea: Papers in Honor of J. Charles Kelley*, ed. Carroll L. Riley and Basil C. Hedrick, 144–164. Carbondale: Southern Illinois University Press.

Smith, A. Ledyard, and A. V. Kidder. 1951. *Excavations at Nebaj, Guatemala.* Washington, DC: Carnegie Institution of Washington, Pub. 594.

Stapert, Dick, and Lykke Johanssen. 1999. "Flint and Pyrite: Making Fire in the Stone Age." *Antiquity* 73: 765–777.

Stuart, David. 2010. "Shining Stones: Observations on the Ritual Meaning on Early Maya Stelae." In *The Place of Stone Monuments: Context, Use and Meaning in Mesoamerica's Preclassic Transition*, ed. Julia Guernsey, John E. Clark, and Barbara Arroyo, 283–298. Washington, DC: Dumbarton Oaks.

Tait, Hugh. 1967. "'The Devil's Looking Glass:' The Magical Speculum of Dr. John Dee." In *Horace Walpole: Writer, Politician and Connoisseur*, ed. Warren Huntington Smith, 195–212. New Haven: Yale University Press.

Taube, Karl. 1992a. "The Iconography of Mirrors at Teotihuacan." In *Art, Polity, and the City of Teotihuacan*, ed. Janet C. Berlo, 169–204. Washington, DC: Dumbarton Oaks.

Taube, Karl. 1992b. *The Major Gods of Ancient Yucatan.* Studies in Precolumbian Art and Archaeology 32. Washington, DC: Dumbarton Oaks.

Taube, Karl. 1994. "The Iconography of Toltec Period Chichen Itza." In *Hidden in the Hills: Maya Archaeology of the Northwestern Yucatan Peninsula*, edited by Hanns J. Prem, 212–246. Acta Mesoamericana 7. Möckmühl: Verlag von Flemming.

Taube, Karl. 2000. "Lightning Celts and Corn Fetishes: The Formative Olmec and the Development of Maize Symbolism in Mesoamerica and the American Southwest." In *Olmec Art and Archaeology: Social Complexity in the Formative Period*, vol. 58. ed. John E. Clark and Mary Pye, 297–337. Studies in the History of Art. Washington, DC: National Gallery of Art.

Taube, Karl. 2001a. "Mirrors." In *The Archaeology of Ancient Mexico and Central America: An Encyclopedia*, ed. Susan Evans and David Webster, 473–474. New York: Garland Publishing.

Taube, Karl. 2001b. "Yahui." In *The Oxford Encyclopedia of Mesoamerican Cultures*, vol. 3. ed. David Carrasco, 359–360. Oxford: Oxford University Press.

Taube, Karl. 2004a. "Flower Mountain: Concepts of Life, Beauty and Paradise Among the Classic Maya." *Res: Anthropology and Aesthetics* 45: 69–98.

Taube, Karl. 2004b. *Olmec Art at Dumbarton Oaks.* Washington, DC: Dumbarton Oaks.

Taube, Karl. 2009. "La religion à Teotihuacan." In *Teotihuacan, cité des Dieux*, by Karl Taube, 152–59. Paris: Musée du quai Branly, Somogy.

Taube, Karl. 2012a. "Jade Maya: Piedra de dioses y reyes antiguos." In *Piedras del cielo: Civilizaciones del jade*, ed. Mariana Roca Cogordan, 33–40. Mexico City: Instituto Nacional de Antropología e Historia.

Taube, Karl. 2012b. "The Symbolism of Turquoise in Postclassic Mexico." In *Turquoise in Mexico and North America: Science, Conservation, Culture and Collections*, edited by J. C. H. King, Max Carocci, Carolyn Cartwright, Colin McEwan, and Rebecca Stacy, 117–134. London: The British Museum.

Taube, Karl, William Saturno, David Stuart, and Heather Hurst. 2010. *The Murals of San Bartolo, El Peten, Guatemala, Part 2: The West Wall.* Ancient America 10. Barnardsville: Center for Ancient American Studies.

Taube, Karl A., Virginia G. Sisson, Russell Seitz, and George E. Harlow. 2004. "The Sourcing of Mesoamerican Jade: Expanded Geological Reconnaissance in the Motagua Region, Guatemala." Appendix to *Olmec Art at Dumbarton Oaks*, by Karl A. Taube, 203–220. Washington, DC: Dumbarton Oaks.

Tedlock, Dennis. 1996. *Popol Vuh: The Definitive Edition of the Mayan Book of the Dawn of Life and the Glories of Gods and Kings.* revised edition. New York: Simon and Schuster.

Villa Córdova, Tómas. 2010. "La cueva y sus reflejos: Los *tezacuitlapilli* de la Pirámide de Sol." *Arqueología* (May–August): 110–135.

Vogt, Evon Z. 1976. *Tortillas for the Gods: A Symbolic Analysis of Zinacanteco Rituals.* Cambridge, MA: Harvard University Press.

Woodbury, Richard W., and Aubrey S. Trik. 1953. *The Ruins of Zaculeu, Guatemala.* Richmond: The William Byrd Press.

Woolley, Benjamin. 2001. *The Queen's Conjurer: The Science and Magic of Dr. John Dee, Advisor to Queen Elizabeth I.* New York: Henry Holt and Company.

Contributors

Marc G. Blainey, Trent University Archaeological Research Centre (TUARC)

Thomas Calligaro, Centre de Recherche et de Restauration des Musées de France (C2RMF)

Carrie L. Dennett, University of Calgary

Emiliano Gallaga M., Escuela de Antropología e Historia del Norte de México (EAHNM), Chihuahua

Julie Gazzola, Dirección de Estudios Arqueológicos/INAH

Sergio Gómez Chávez, Zona Arqueológica de Teotihuacán/INAH

Olivia Kindl, El Colegio de San Luis

Brigitte Kovacevich, University of Central Florida

Achim Lelgemann, Universidad Autónoma de San Luis Potosí

José J. Lunazzi, State University of Campinas

John J. McGraw, California State University, Northridge

Emiliano Melgar, Museo del Templo Mayor

Joseph B. Mountjoy, Centro INAH Jalisco

Reyna Solis, Museo del Templo Mayor

Karl Taube, University of California, Riverside

Index